INTRODUCTORY PHYSICAL GEOGRAPHY

INTRODUCTORY PHYSICAL GEOGRAPHY

Kelly Bennett
Brenda Matthews
Samantha Fisher

Kruger Brentt
Publishers
2024

Kruger Brentt Publishers UK. LTD.
Company Number 9728962

Regd. Office: 68 St Margarets Road, Edgware, Middlesex HA8 9UU

ISBN: 978-1-78-7151-74-1

For information on all our publications visit our website at http://krugerbrentt.com/

PREFACE

Physical geography is the study of the Earth's natural systems and processes. It is concerned with the study of natural phenomena, such as climate, vegetation, and landforms. Physical geography is an important field of study, as it helps us to understand the Earth's environment and how it is changing. Physical geography is concerned with the study of the Earth's systems. These systems include the atmosphere, hydrosphere, lithosphere, and biosphere. Each system interacts with the others, and changes in one system can affect the others. Understanding these systems is important for predicting and mitigating natural disasters, such as floods, droughts, and hurricanes. Physical geography is concerned with the study of climate. This includes understanding the factors that influence climate, such as solar radiation, atmospheric circulation, and ocean currents. It also includes studying the impact of climate change on the Earth's environment and its inhabitants. Understanding climate is important for predicting weather patterns and understanding the impacts of climate change.

Physical geography is concerned with the study of landforms. This includes understanding how landforms are created, such as through erosion, glaciation, and tectonic activity. It also includes studying the impact of landforms on human and animal habitats. Understanding landforms is important for predicting natural hazards, such as landslides and earthquakes, and for planning land use Physical geography is concerned with the study of vegetation. This includes understanding the factors that influence plant growth, such as climate, soil, and topography. It also includes studying the impact of vegetation on the environment and human society. Understanding vegetation is important for managing natural resources, such as forests and agricultural land. Physical geography is an important field of study, as it helps us to understand the Earth's environment and how it is changing. By studying the Earth's systems, climate, landforms, and vegetation, physical geographers can predict natural disasters, manage natural resources, and plan land use. As the Earth's environment continues to change, the study of physical geography will become increasingly important for understanding and mitigating the impacts of climate change.

The present book contains 19 chapters namely introduction to geography as a discipline; earth's place within the cosmos; introduction to geology & geologic time; mapping earth's surface; earth-sun relationships: reasons for the seasons; earth's atmosphere; elements of weather and climate; basic mineral development; igneous rocks; sedimentary rocks; metamorphism and metamorphic rocks; weathering and soils; earths dynamic surface: plate tectonics; earths dynamic surface: tectonics force; earths dynamic surface: volcanoes; shaped bycoastal processes; shaped by rivers and running water; shaped by wind as a geomorphic agent & shaped byglaciers. This book is intended to be an introduction to Earth's grid system, rivers, oceans, deserts, basic geology, and cartography. The subject matter is presented in a lucid and succinct style so that even those who do not have a prerequisite knowledge of the subject can easily understand it.

Extensive use of Maps for better clarity. It is widely read reference book that deals with the relevant features and topics of geographical landscape in a systematic and comprehensive manner. It is being referred to by the graduates and postgraduates of the subject and by the researchers and academicians.

We are grateful to all those persons as well as various books, manuals, periodicals, magazines, journals etc. that helped in the preparation of this book. In spite of the best efforts, it is possible that some errors may have occurred into the compilation and editing of the book. Further queries, constructive suggestions and criticisms for the improvement of the book are always welcomed and shall be thankfully acknowledged.

Kelly Bennett

Brenda Matthews

Samantha Fisher

CONTENTS

Chapter 1

INTRODUCTION TO GEOGRAPHY AS A DISCIPLINE

Figure 1.1 Lake Sabrina, a glacially formed feature near Bishop, California. Image by Jeremy Patrich is under a CC BY 4.0 license.

Goals & Objectives of this unit

- Develop an understanding of geographic and scientific knowledge and inquiry.
- Describe the basic model of the scientific method and how scientists use it to understand the natural world.
- Explain the importance of understanding location, including the latitude & longitude.
- Compare and contrast the various types of geospatial technologies used today.

SCIENTIFIC & GEOGRAPHIC INQUIRY

Physical Geography is the study of our home planet and all of its components: its lands (lithosphere), waters (hydrosphere), living organisms (biosphere), atmosphere, and interior. In this book, some chapters are devoted to the processes that shape the lands and impact people. Other chapters depict the processes of the atmosphere and its relationship to the planet's surface and all our living creatures. For as long as people have been on the planet, humans have had to live within Earth's boundaries. Now human life is having a profound effect on the planet, with both a positive and negative result. The journey to better understanding Earth begins here with an exploration of how scientists learn about the natural world and introduce you to the study of physical geography.

Scientific Inquiry

Science is a path to gaining knowledge about the natural world. The study of science also includes the body of knowledge that has been collected through scientific inquiry. To conduct a scientific investigation, scientists ask testable questions that can be systematically observed and carefully evidenced collected. Then they use logical reasoning and some imagination to develop a testable idea, called a hypothesis, along with explanations to explain the idea. Finally, scientists design and conduct experiments based on their hypotheses.

Scientists seek to understand the natural world by asking questions and then trying to answer the questions with evidence and logic. A scientific question must be testable and supported by empirical data, it does not rely on faith or opinion. Our understanding of Earth's natural processes helps us to answer questions such as, why earthquakes occur where they do and what are the consequences of adding excess greenhouse gases into the atmosphere.

Scientific research may be done to build knowledge or to solve problems, and lead to scientific discoveries and technological advances. Research often aids in the development of applied research. Sometimes the results of the research may be applied long after the research was completed. Sometimes the results are discovered while scientists are conducting their research. Some ideas are not testable. For example, supernatural phenomena, such as stories of ghosts, vampires, or The Yeti, cannot be tested. Scientists describe what they see, whether in nature or a laboratory.

Science is the realm of facts and observations, not moral judgments. Scientists increase our technological knowledge, but science does not determine how or if we use that knowledge. Scientists learned to build an atomic bomb, but scientists didn't decide whether or when to use it. Scientists have accumulated data on warming temperatures; their models have shown the likely causes of this warming.

But although scientists are largely in agreement on the causes of global warming, they can't force politicians or individuals to pass laws, or change behaviors.

For science to work, scientists must make some assumptions. The rules of nature, whether simple or complex, are the same everywhere in the universe. Natural events, structures, and landforms have causes and evidence from the world can be used to learn about those causes. The objects and events in nature can be understood through careful, systematic study.

Scientific ideas can change if we gather new data or learn more. An idea, even one thatis accepted today, may need to be changed slightly or be entirely replaced if new evidence is found that contradicts it. Scientific knowledge can withstand the test of time because accepted ideas in science become more reliable as they survive more tests.

Geographic Inquiry

- Geography is the study of the physical and cultural environments of the earth. What makes geography different from other disciplines is its focus on spatial inquiry and analysis.
- Geographers also try to look for connections between things such as patterns, movement and migration, trends, and so forth. This process is called either geographic or spatial inquiry.
- To do this, geographers go through a geographic methodology that is quite similar to the scientific method, but again with a geographic or spatial emphasis.
    1. Ask a geographic question. Ask questions about spatial relationships in the world around you, such as the location of your college as it pertains to your home, high school, or work.
    2. Acquire geographic resources. Identify data and information that you need to answer your question.
    3. Explore geographic data. Turn the data into maps, tables, and graphs, and look for patterns and relationships by utilizing geospatial computer programs and statistics.
    4. Analyze geographic information. Determine what the patterns and relationships mean concerning your question. This is where critical thinking comes to play; once observing the results you then begin to develop future work or perhaps ask even more questions.

"Knowing where something is, how its location influences its characteristics, and how its location influences relationships with other phenomena are the foundation of geographic thinking. Like other research methods, it also asks you to explore, analyze, and act upon the things you find. It also is important to recognize that this is the same method used by professionals around the world working to address social, economic, political, environmental and a range of scientific issues." (ESRI)

THE SCIENTIFIC METHOD

You have probably learned that the scientific method is a series of steps that help answer research questions. Scientists use data and evidence gathered from observations, experience, or experiments to answer their questions and to essentially create even more hypotheses.

But scientific inquiry rarely proceeds in the same sequence of steps outlined by the scientific method. For example, the order of the steps might change because more questions arise from the data that is collected. Still, to come to verifiable conclusions, logical, repeatable steps of the scientific method must be followed. An example flow chart has been provided below to show the steps of the scientific method. As an example, what if your question was, 'do taller people wear bigger shoes'? What data would you need? How would you collect that data, and how would you test your hypothesis?

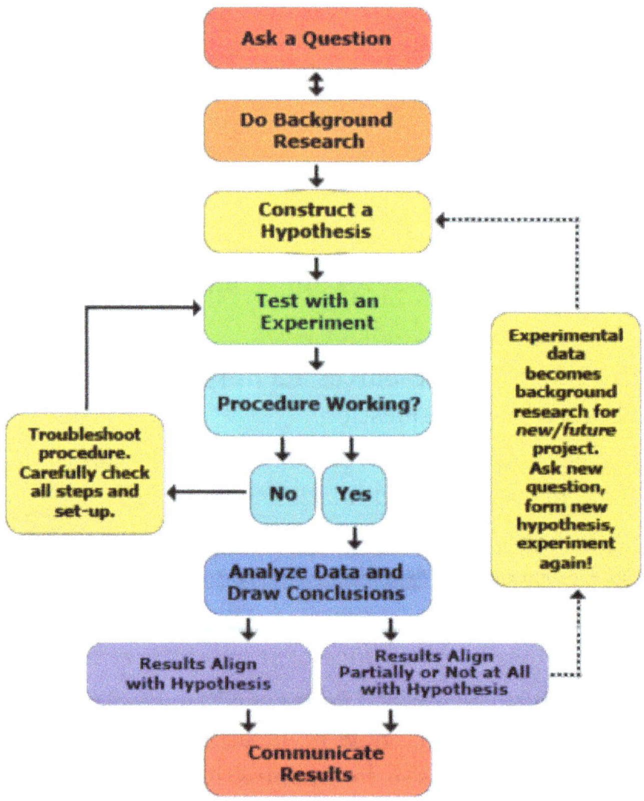

Figure 1.2 Flow chart of the Scientific Method. Image is in the public domain.

Scientific Questioning

- The most important thing a scientist can do is to ask critical thinking questions.
- What makes the San Andreas Fault different from the Hollywood Fault?
- Why does Earth have so many varied life forms but other local planets in the solar system do not?
- What impacts could a warmer planet have on weather and climate systems?

Geographers can answer testable questions about the natural world, but what makes a question impossible to test? Some untestable questions are whether ghosts exist or whether there is life after death. A testable question might be about how to reduce soil erosion on a farm. A farmer has heard of a planting method called "no-till farming." Using this process eliminates the need for plowing the land. The farmer's question is: Will no-till farming reduce the erosion of the farmland?

Scientific research

To answer a question, a scientist first finds out what is already known about the topic by reading books and magazines, searching the Internet, and talking to experts. This information will allow the scientist to create a good experimental design. If this question has already been answered, the research may be enough, or it may lead to new questions.

Figure 1.3 Farmers working in a greenhouse. Image by USDA.gov in the public domain.

Example: The farmer researches no-till farming on the Internet, at the library, at the local farming supply store, and a few local farmers in his area. He learns about various farming methods. He learns what type of fertilizer is best to use and what the best crop spacing would be. From his research, he learns that no-till farming can be a way to reduce carbon dioxide emissions into the atmosphere, which helps in the fight against global warming.

Hypothesis

With the information collected from background research, the scientist creates a plausible explanation for the question. This is a hypothesis. The hypothesis must directly relate to the question and must be testable. Having a hypothesis guides a scientist in designing experiments and interpreting data. Example: The farmer's hypothesis is this: No-till farming will decrease soil erosion on hills of similar steepness as compared to the traditional farming technique because there will be fewer disturbances to the soil.

Data Collection: Observation & Experimentation

To support or refute a hypothesis, the scientist must collect data. A great deal of logic and effort goes into designing tests to collect data so the data can answer scientific questions. Data is usually collected by experiment or observation.

Figure 1.4 Example of no-till farming in a soybean field. Image by Tim McCabe, USDA Natural Resources Conservation Service is in the public domain.

Observation is used to collect data when it is not possible, for practical or ethical reasons, to perform experiments. Written descriptions are examples of qualitative data based on observations. Scientists use many different types of instruments to make quantitative measurements, such as an electron microscope can be used to explore tiny objects or telescopes to learn about the universe.

Experiments may involve chemicals and test tubes, or they may require advanced technologies like a high-powered electron microscope or radio telescope. Atmospheric scientists may collect data by analyzing the gases present in gas samples, and geochemists may perform chemical analyses on rock samples. A good experiment must have one factor that can be manipulated or changed. This is the independent variable. The rest of the factors must remain the same. They are the experimental controls. The outcome of the experiment, or what changes as a result of the experiment, is the dependent variable. The dependent variable "depends" on the independent variable.

As an example: The farmer experiments on two separate hills. The hills have similar steepness and receive similar amounts of sunshine. On one, the farmer uses a traditional farming technique that includes plowing. On the other, he uses a no-till technique, spacing plants farther apart and using specialized equipment for planting. The plants on both hillsides receive identical amounts of water and fertilizer. The farmer measures plant growth on both hillsides. In this experiment: what are the independent, experimental and dependent variables? The independent variable is the farming technique—either traditional or no-till—because that is what is being manipulated. For a fair comparison of the two farming techniques, the two hills must have the same slope and the same amount of fertilizer and water. These are the experimental controls. The amount of erosion is the dependent variable. It is what the farmer is measuring. During an experiment, scientists make many measurements. Data in the form of numbers is quantitative.

Data gathered from advanced equipment usually goes directly into a computer, or the scientist may put the data into a spreadsheet. Charts and tables display data and should be clearly labeled. Statistical analysis makes more effective use of data by allowing scientists to show relationships between different categories of data. Statistics can make sense of the variability in a data set. Graphs help scientists to visually understand the relationships between data.Pictures are created so that other interested people can see the relationships easily.

Conclusions

Scientists study graphs, tables, diagrams, images, descriptions, and all other available data to conclude from their experiments. Is there an answer to the question based on the results of the experiment? Was the hypothesis supported? Some experiments completely support a hypothesis, and some do not. If a hypothesis is shown to be wrong, the experiment was not a failure. All experimental results contribute to knowledge. Experiments that do or donot support a hypothesis may lead to even more questions and more experiments.

Example: After a year, the farmer finds that erosion on the traditionally farmed hill is 2.2 times greater than erosion on the no-till hill. The plants on the no-till plots are taller and the soil moisture is higher. The farmer decides to convert to no-till farming for future crops. The farmer continues researching to see what other factors may help reduce erosion.

Theory

As scientists conduct experiments and make observations to test a hypothesis, over time they collect a lot of data. If a hypothesis explains all the data and none of the data contradicts the hypothesis, the hypothesis becomes a theory. A scientific theory is supported by many observations and has no major inconsistencies. A theory must be constantly tested and revised. Once a theory has been developed, it can be used to predict behavior. A theory provides a model of reality that is simpler than the phenomenon itself. Even a theory can be overthrown if conflicting data is discovered. However, a longstanding theory that has lots of evidence to back it up is less likely to be overthrown than a newer theory.

Science does not prove anything beyond a shadow of a doubt. Scientists seek evidence that supports or refutes an idea. If there is no significant evidence to refute an idea and a lot of evidence to support it, the idea is accepted. The more lines of evidence that support an idea, the more likely it will stand the test of time. The value of a theory is when scientists can use it to offer reliable explanations and make accurate predictions.

GEOGRAPHIC GRID SYSTEM

Geography is about spatial understanding, which requires an accurate grid system to determine absolute and relative location. Absolute location is the exact x- and y- coordinate on the Earth. Relative location is the location of something relative to other entities. For example, when you use Google Maps, you put in an absolute location. But as you start driving, the device tells you to turn right or left relative to objects on the ground: “Turn left on exit Valencia Blvd” is relative to the other exit points. Or if you give directions to your house, you often use relative locations to help them understand how to get to your house.

Great & Small Circles

Much of Earth’s grid system is based on the location of the North Pole, South Pole, and the Equator. The poles are considered points. The plane of the equator is an imaginary horizontal line that cuts the earth into two equal halves. This brings up the topic of great and small circles. A great circle is any circle that divides the earth into a circumference of two equal halves. It’s also the largest circle that can be drawn on a sphere. The line connecting any points along a great circle is also the shortest distance between those two points. Examples of great circles include the Equator, all lines of longitude, the line that divides the earth into day and night called the circle of illumination, and the plane of the ecliptic, which divides the earth into equal halves along the equator. Small circles are circles that cut the earth, but not into equal halves.

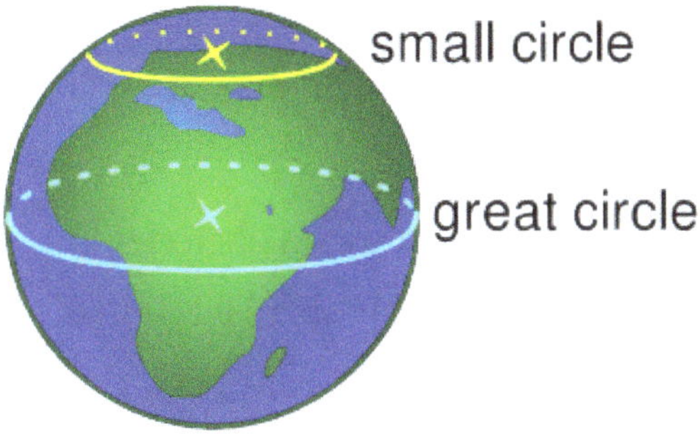

Figure 1.5 Great & small circles. Image by Brian Brondel is under a CC BY-SA 2.5 license.

Latitude & Longitude

Many assume that latitude is a line connecting points on the earth and it's not. Latitude is an angular measurement north or south of the equator. So, 30 degrees north means a point that is 30 degrees north of the equator. Latitude is also expressed in degrees, minutes, and seconds; 360 degrees in a circle, 60 minutes (') in a degree, and 60 seconds (") in a minute. When you use Google Earth, the coordinate locations are in this degrees/minutes/second's format.

Latitude varies from 0 degrees (equator) to 90 degrees north and south (the poles).

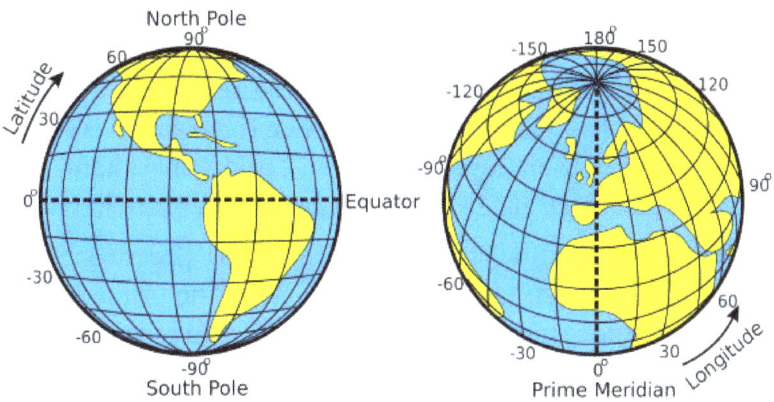

Figure 1.6 Latitude & Longitude. Image by Djexplo has been designated to the public domain under a CC0 1.0 Universal Public Domain Dedication

A line connecting all points of the same latitude is called a parallel, because the lines run parallel to each other. The only parallel that is also a great circle is the equator. All other parallels are small circles. The following are the most important parallel lines:

- Tropic of Cancer, 23.5 degrees N
- Tropic of Capricorn, 23.5 degrees S
- Arctic Circle, 66.5 degrees N
- Antarctic Circle, 66.5 degrees S
- North Pole, 90 degrees N (infinitely small circle)
- South Pole, 90 degrees S (infinitely small circle)

Latitude is also sometimes described as zones of latitude. Some of these zones of latitude include:

- Low latitude - generally between the equator and 30 degrees N
- Midlatitude - between 30 degrees and 60 degrees N and S
- High latitude - latitudes greater than about 60 degrees N and S
- Equatorial - within a few degrees of the equator
- Tropical - within the tropics (between 23.5 degrees N and 23.5 degrees S
- Subtropical - slightly pole-ward of the tropics, generally around 25-30 degrees N and S
- Polar - within a few degrees of the North or South Pole

Longitude is the angular measurement east and west of the Prime Meridian. Like latitude, longitude is measured in degrees, minutes, and seconds. Lines connecting equal points of longitude are called meridians. But unlike parallels, meridians do not run parallel to each other. Rather they are farthest apart from each other at the equator and merge toward each other toward the poles. The problem with longitude is that there isn't a natural baseline like the equator is for latitude. For over a hundred years, nations used their own "prime meridian" which proved problematic for trade. But in 1883 an international conference in Washington.

After weeks of debate, the Royal Observatory at Greenwich, England was determined as the Greenwich Meridian or also called the prime meridian for the world. So today, longitude starts at the Prime Meridian and measures east and west of that line.

At 180 degrees of the Prime Meridian in the Pacific Ocean is the International Date Line. The line determines where the new day begins in the world. Now because of this, the International Date Line is not straight, rather it follows national borders so that a country isn't divided into two separate days (and we think our time zones are a pain). If you look at the map on the next page, the International Date Line is to the right in a dark, black line. Note how it is drawn to make sure nations are not divided by the International Date Line.

Time Zones

This is also a good time to take a look at time zones around the world. If you refer to the map on the next page, you can see the different time zones in the various colors. Since the earth rotates 360 degrees in a 24-hour period, the earth rotates 15 degrees every hour creating 24 time zones. In an ideal world, each time zone would follow lines of longitude every 15 degrees (7.5° in each direction from the center of the time zone). But because of political boundaries, time zones are not divided up so perfectly and vary greatly in shape and width.

Greenwich, England was chosen in the mid-nineteenth century as the starting point of time worldwide. The reason was that at the time, England was the superpower of the time both militarily and economically. So, the meridian that ran through Greenwich became zero degrees or the prime meridian. Because of the earth's rotation in reference to the prime meridian, locations east of the new meridian meant time was ahead while locations west of the meridian were behind in time in reference to Greenwich, England. Ultimately, when you combine parallel and meridian lines, you end up with a geographic grid system that allows you to determine your exact location on the planet.

Pin It! Time Zones

Visit this interactive Time Zone Map for more information on time zones.

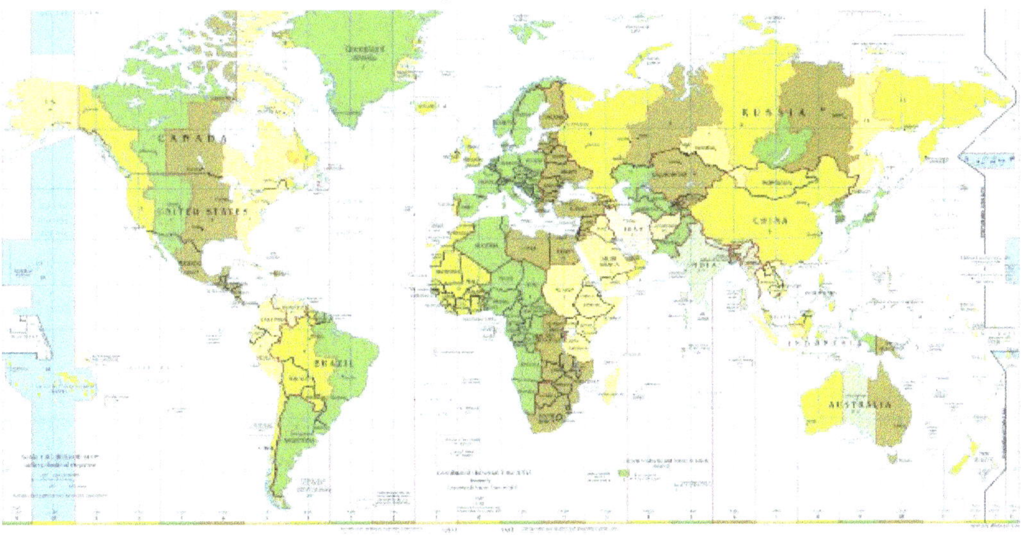

Figure 1.7 Global map showing time zone distribution. Image by Mr. Mingsz is in the public domain.

GEOSPATIAL TECHNOLOGY

Data, data, data... data is everywhere. It's collected every time you go to the grocery store and use their card to reduce the costs when you click on a link on Facebook, or when you do any kind of search on a search engine like Google, Bing, or Yahoo!. It is used by the state department of transportation when you are driving on a freeway, or when you use an app on a smartphone. Futurists believe that in the near future, face recognition technology will allow a sales representative to know what types of clothes you like to buy based on a database of your recent purchases at their store and others.

Now there are two basic types of data you need to know: spatial and non-spatial data. Spatial data, also called geospatial data, is data that can be linked to a specific location on Earth.

Geospatial data is becoming "big business" because it isn't just data, but data that canbe located, tracked, patterned, and modeled based on other geospatial data. Census information that is collected every 10 years is an example of spatial data. Non-spatial data is data that cannot be specifically traced to a specific location. This might include the number of people living in a household, enrollment within a specific course, or gender information. But non-spatial data can easily become spatial data if it can be linked in some way to a location.

Geospatial technology specialists have a method called geocoding that can be used to give non- spatial data a geographic location. Once data has a spatial component associated with it, the type of questions that can be asked dramatically changes.

Remote Sensing

Remote sensing can be defined as the ability to study objects without being in direct physical contact with them. For example, your eyes are a form of passive remotesensing because they are "passively" absorbing electromagnetic energy within the visible spectrum from distant objects and your brain is processing that energy into information.

There are a variety of remote sensing platforms or devices, but they can be categorized into the following that we will look at throughout the course. Satellite imagery is a type of remotely sensed imagery taken of the Earth's surface, which is produced from orbiting satellites that gather data via electromagnetic energy. Next is aerial photography, which is film-based or digital photographs of the Earth, usually from an airplane or non-piloted drone. Images are either taken from a vertical or oblique position. The third is radar, which is an interesting form of remote sensing technology that uses microwave pulses to create imagery of features on Earth. This can be from a satellite image or ground-based Doppler radar for weather forecasting. Finally, a fast-growing realm of remote sensing is called Light Detection and Ranging or Lidar, which is a form of remote sensing that measures the distance of objects using laser pulses of light.

Figure 1.8 Remote sensing of the environment. Used with permission from gisgeography.com

Global Positioning Systems

Another type of geospatial technology, and a key technology for acquiring accurate control points on Earth's surface, is global positioning systems (GPS). In order to determine the location of a GPS receiver on Earth's surface, a minimum of four satellites are required using a mathematical process called triangulation. Normally the process of triangulation requires a minimum of three transmitters, but because the energy sent from the satellite is traveling at the speed of light, minor errors in calculation could result in large location errors on the ground. Thus, a minimum of four satellites is often used to reduce this error. This process using the geometry of triangles to determine location is used not only in GPS but a variety of other location needs, like finding the epicenter of earthquakes.

A user can use a GPS receiver to determine their location on Earth through a dynamic conversation with satellites in space. Each satellite transmits orbital information called the ephemeris using a highly accurate atomic clock along with its orbital position called the almanac. The receiver will use this information to determine its distance from a single satellite using the equation D = rt, where D = distance, r = rate or the speed of light (299,792,458 meters per second), and t = time using the atomic clock.

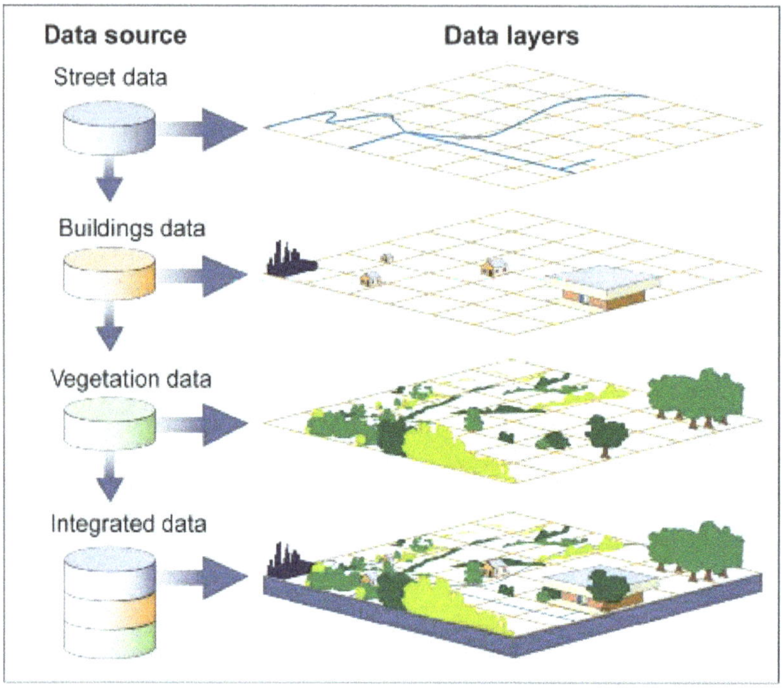

Figure 1.9 Visual Representation of Themes in a GIS. Image is in the public domain.

There is a technology that is capable of bringing together remote sensing data, GPS data points, spatial and non-spatial data, and spatial statistics into a single, dynamic system for analysis, and that is a geographic information system (GIS). A GIS is a powerful database system that allows users to acquire, organize, store, and most importantly analyze information about the physical and cultural environments. A GIS views the world as overlaying physical or cultural layers, each with quantifiable data that can be analyzed. A single GIS map of a national forest could have layers such as elevation, deciduous trees, evergreens, soil type, soil erosion rates, rivers and tributaries, major and minor roads, forest health, burn areas, regrowth, restoration, animal species type, trails, and more. Each of these layers would contain a database of information specific to that layer.

Nearly every discipline, career path, or academic pursuit uses geographic information systems because of the vast amount of data and information about the physical and culturalworld. Disciplines and career paths that use GIS include conservation, ecology, disaster response and mitigation, business, marketing, engineering, sociology, demography, astronomy, transportation, health, criminal justice and law enforcement, travel and tourism, news media, and the list could endlessly go on.

Now, GIS primarily works from two different spatial models: raster and vector. Raster based GIS models are images much like a digital picture. Each image is broken down into a series of columns and rows of pixels and each pixel is georeferenced to somewhere on Earth's surface is represents a specific numeric value - usually a specific color or wavelength withinthe electromagnetic spectrum. Most remote sensing images come into a GIS as a raster layer. The other type of GIS model is called a vector model. Vector-based GIS models are based on the concept of points that are again georeferenced (e.g. given an x-, y-, and possibly z- location) to somewhere specific on the ground. From points, lines can be created by connecting a series of points and areas can be created by closing loops of vector lines. For each of these vector layers, a database of information can be attributed to it. As an example, a vector line of rivers could have a database associated with it such as length, width, streamflow, government agencies responsible for it, and anything else the GIS user wants to tie to it. What these vector models represent is also a matter of scale. For example, a city can be represented as a point or a polygon depending on how zoomed in you are to the location. A map of the world would show cities as points, whereas a map of a single county may show the city as a polygon with roads, populations, pipes, or grid systems within it.

SUMMARY

Physical geography is the spatial study of our home planet and all of its components: its lands, waters, atmosphere, and interior. Like other sciences, physical geography is a science that is grounded in scientific knowledge using the scientific method as the fundamental way to understand the environment.

Geographers and all spatial scientists require a strong background in understanding the way humans have partitioned the Earth to determine location. In order to do that, a series of lines representing angular measurements on the earth was established, known as the geographic grid system. Once that has been done, spatial knowledge can be collected and analyzed based on geographic or spatial data. This allows us to understand spatial concepts of patterns, distributions and flows based on location and spatial boundaries.

Often times this geographic data must be collected and analyzed using a high-tech and dynamic technology called geospatial technology. This technology encompasses powerful remote sensing technology, global positioning systems, and geographic information systems.

2 Chapter

EARTH'S PLACE WITHIN THE COSMOS

Figure 2.1 The Milky Way—Our Galaxy. Image by NASA is in the public domain.

Goals & Objectives of this unit

- Understand the scientific ideas of how the universe formed and is expanding.
- Compare and contrast the difference and similarities between dark matter and dark energy.
- Describe star systems and the various types of galaxies.
- Explain the phenomenal power within stars.
- Classifying and measuring distant stars.

INTRODUCTION TO THE UNIVERSE

The Whirlpool Galaxy, also known as M51, is a spiral galaxy about 23 million light-years from Earth. Its interactions with the yellowish dwarf galaxy NGC 5195 are of interest to astronomers because the galaxies are near enough to Earth to be well-studied. Decades ago, astronomers could not tell if these two galaxies were just passing each other but radio astronomy has supplied astronomers with important data outlining their interactions. Using this data, astronomers have simulated the interaction. NGC 5195 came from behind and then passed through the main disk of M51 about 500 to 600 million years ago. The dwarf galaxy crossed the disk again between 50 and 100 million years ago and is now slightly behind M51. These interactions appear to have intensified the spiral arms that are the dominant characteristic of the Whirlpool Galaxy.

Astronomers can learn about objects unimaginably far away from Earth using telescopes that sense all wavelengths on the electromagnetic spectrum. Imagine what Galileo would do if he could see the images and data astronomers have available to them now. The study of the universe is called cosmology. Cosmologists study the structure and changes in the present universe. The universe contains all of the star systems, galaxies, gas, and dust, plus all the matter and energy that exists now, that existed in the past, and that will exist in the future. The universe includes all of space and time.

EXPANDING UNIVERSE

What did the ancient Greeks recognize as the universe? In their model, the universe contained Earth at the center, the Sun, the Moon, five planets, and a sphere to which all the stars were attached. This idea held for many centuries until Galileo's telescope helped allow people to realize that Earth is not the center of the universe. They also found out that there are many more stars than were visible to the naked eye. All of those stars were in the Milky Way Galaxy. In the early 20th century, an astronomer named Edwin Hubble discovered that what scientists called the Andromeda Nebula was over two million light-years away, many times farther than the farthest distances that had ever been measured. Hubble realized that many of the objects that astronomers called nebulae were not clouds of gas but were collections of millions or billions of stars that we now call galaxies.

Hubble showed that the universe was much larger than our galaxy. Today, we know that the universe contains about a hundred billion galaxies, about the same number of galaxies as there are stars in the Milky Way Galaxy. After discovering that there are galaxies beyond the Milky Way, Edwin Hubble went on to measure the distance to hundreds of other galaxies. His data would eventually show how the universe is changing and would even yield clues as to how the universe formed. Today we know that the universe is nearly 14 billion years old.

Redshift

If you look at a star through a prism, you will see a spectrum or a range of colors through the rainbow. The spectrum will have specific dark bands where elements in the star absorb light of certain energies. By examining the arrangement of these dark absorption lines, astronomers can determine the composition of elements that make up a distant star. The element helium was first discovered in our Sun, not on Earth, by analyzing the absorption lines in the spectrum of the Sun.

While studying the spectrum of light from distant galaxies, astronomers noticed something strange. The dark lines in the spectrum were in the patterns they expected, but they were shifted toward the red end of the spectrum, as shown in Figure below. This shift of absorption bands toward the red end of the spectrum is known as redshift.

Redshift occurs when the light source is moving away from the observer or when the space between the observer and the source is stretched. What does it mean that stars and galaxies are redshifted? When astronomers see redshift in the light from a galaxy, they know that the galaxy is moving away from Earth. What astronomers are noticing is that all the galaxies have a redshift, strongly indicating that all galaxies are moving away from each other causing the Universe to expand.

Redshift can occur with other types of waves too, called the Doppler Effect. An analogy to redshift is the noise a siren makes as it passes you. You may have noticed that an ambulance seems to lower the pitch of its siren after it passes you. The sound waves shift towards a lower pitch when the ambulance speeds away from you. Though redshift involves light instead of sound, a similar principle operates in both situations.

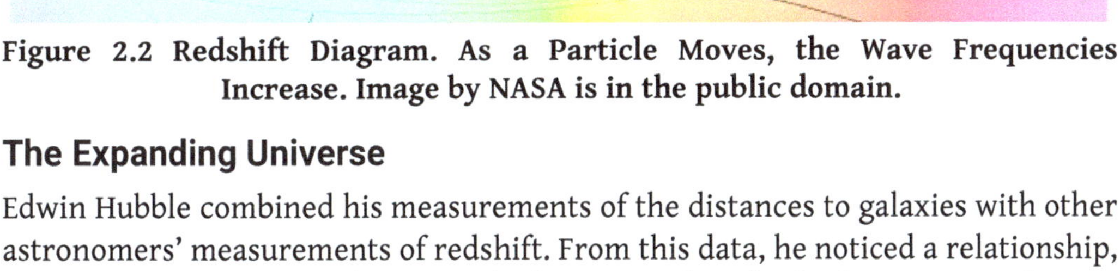

Figure 2.2 Redshift Diagram. As a Particle Moves, the Wave Frequencies Increase. Image by NASA is in the public domain.

The Expanding Universe

Edwin Hubble combined his measurements of the distances to galaxies with other astronomers' measurements of redshift. From this data, he noticed a relationship, which is now called Hubble's Law. The law states that the farther away a galaxy is, the faster it is moving away from us. What this leads to is the hypothesis that the universe is expanding. The figure below by NASA shows a simplified diagram of the expansion of the universe. If you look closely at the diagram, it is observed that on the left was the formation of the universe, and the energy is quite high. Over the course of the 13.7 billion years, the energy begins to cool enough to create trillions of stars and over time develop into galaxies. Over time, the galaxies continue to cool and expand farther apart from each other.

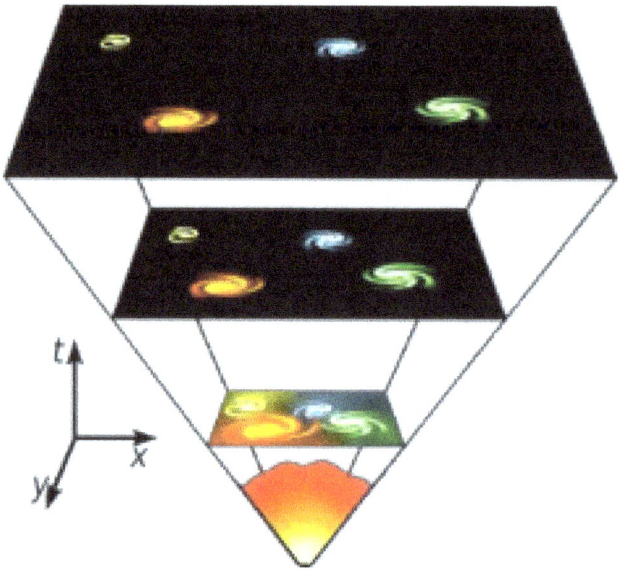

Figure 2.3 Shows Slices of Expansion of Universe Without an Initial Singularity. Image by NASA is in the public domain.

FORMATION OF THE UNIVERSE

Before Hubble, most astronomers thought that the universe didn't change. But if the universe is expanding, what does that say about where it was in the past? If the universe is expanding, the next logical thought is that in the past it had to have been smaller.

The Big Bang Theory

Big Bang Theory is the most widely accepted cosmological explanation of how the universe formed. If we start at the present and go back into the past, the universe is contracting, getting smaller and smaller. What is the result of a contracting universe? According to the Big Bang theory, the universe began about 13.7 billion years ago. Everything that is now in the universe was squeezed into a very small volume. Imagine all of the known universes in a single, hot, chaotic mass. An enormous explosion, a big bang, caused the universe to start expanding rapidly. All the matter and energy in the universe, and even space itself, came out of thisexplosion. What came before the Big Bang? There is no way for scientists to know since there is no remaining evidence.

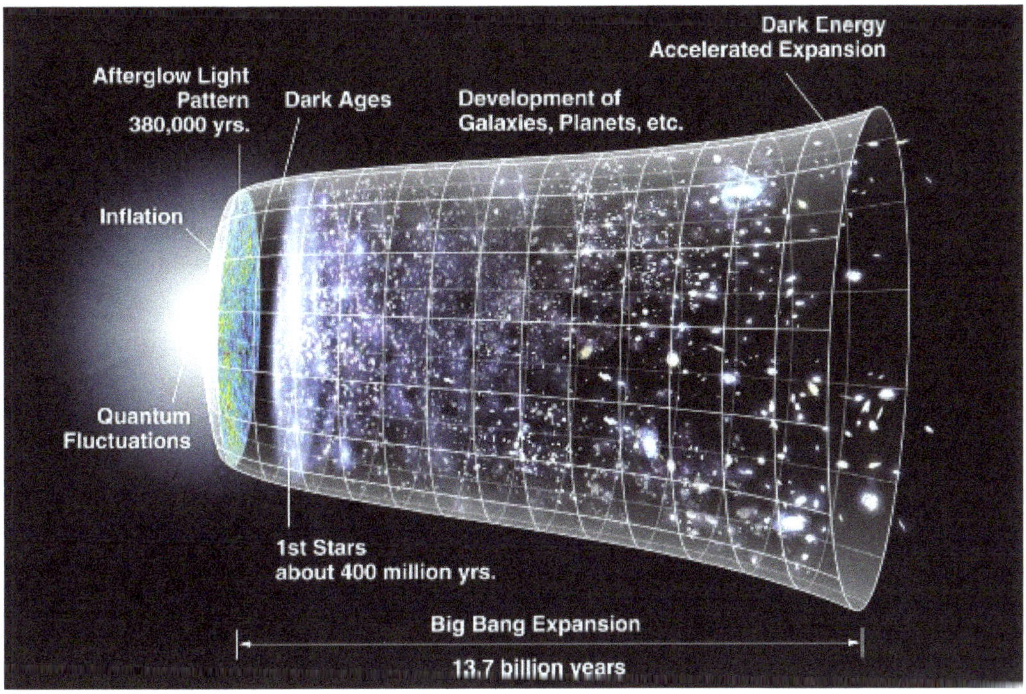

Figure 2.4 The Big Bang Theory, Measuring the Expansion Over a Period of 13.7 Billion Years. Image by NASA is in the public domain.

After the Big Bang

In the first few moments after the Big Bang, the universe was unimaginably hot and dense. As the universe expanded, it became less dense and began to cool. After only a few seconds, protons, neutrons, and electrons could form. After a few minutes, those subatomic particles came together to create hydrogen. The energy in the universe was great enough to initiate nuclear fusion and hydrogen nuclei were fused into helium nuclei. The first neutral atoms that included electrons did not form until about 380,000 years later. The matter in the early universe was not smoothly distributed across space. Dense clumps of matter held close together by gravity were spread around. Eventually, these clumps formed countless trillions of stars, billions of galaxies, and other structures that now form most of the visible mass of the universe. If you look at an image of galaxies at the far edge of what we can see, you are looking at great distances. But you are also looking across a different type of distance. What do those far away galaxies represent? Because it takes so long for light from so far away to reach us, you are also looking back in time.

Dark Matter

The Big Bang Theory is still the best scientific model we have for explaining the formation of the universe and many lines of evidence support it. However, recent discoveries continue to shake up our understanding of the universe. Astronomers and other scientists are now wrestling with some unanswered questions about what the universe is made of and why it is expanding. A lot of what cosmologists do is create mathematical models and computer simulations to account for these unknown phenomena, such as dark energy and dark matter.

Scientists are much more certain what dark matter is not than we are what it is. First, it is dark, meaning that it is not in the form of stars and planets that we see. Observations show that there is far too little visible matter in the universe to make up the 27% required by the observations. Second, it is not in the form of dark clouds of normal matter, matter made up of particles called baryons. We know this because we would be able to detect baryonic clouds by their absorption of radiation passing through them. Third, dark matter is not antimatter, because we do not see the unique gamma rays that are produced when antimatter annihilates with matter. Finally, we can rule out large galaxy-sized black holes based on how many gravitational lenses we see. High concentrations of matter bend light passing near them from objects further away, but we do not see enough lensing events to suggest that such objects make up the required 25% dark matter contribution.

Dark Energy

Astronomers who study the expansion of the universe are interested in knowing the rate of that expansion. Is the rate fast enough to overcome the attractive pull of gravity?

- If yes, then the universe will expand forever, although the expansion will slow down
- over time.
- If no, then the universe would someday start to contract, and eventually get squeezed together in a big crunch, the opposite of the Big Bang.

Recently astronomers have made a discovery that answers that question: the rate at which the universe is expanding is increasing. In other words, the universe is expanding faster now than ever before, and in the future, it will expand even faster. So now astronomers think that the universe will keep expanding forever. But it also proposes a perplexing new question: What is causing the expansion of the universe to accelerate? One possible hypothesis involves a new, hypothetical form of energy called dark energy. Some scientists think that dark energy makes up as much as 72% of the total energy content of the universe.

STAR SYSTEMS & GALAXIES

Although constellations have stars that usually only appear to be close together, stars may be found in the same portion of space. Stars that are grouped closely together are called star systems. Larger groups of hundreds or thousands of stars are called star clusters. The image shown here is a famous star cluster classed M45, also known as Pleides, which can be seen with the naked autumn sky. Although the star humans know best is a single star, many stars, more than half of the bright stars in our galaxy are star systems. A system of two stars orbiting each other is a binary star. A system with more than two stars orbiting each other is a multiple star system. The stars in a binary or multiple star system are often so close together that they appear as only through a telescope can the pair be distinguished.

Figure 2.5 Star Cluster M45- Pleides. Image by NASA is in the public domain.

Star Systems

Star clusters are divided into two main types, open clusters, and globular clusters. Open clusters are groups of up to a few thousand stars that are loosely held together by gravity. The Pleiades is an open cluster that is also called the Seven Sisters. Open clusters tend to be blue and often contain glowing gas and dust. Open clusters are made of young stars that are formed from the same nebula. The stars may eventually be pulled apart by gravitational attraction to other objects.

Globular Clusters

Globular clusters are groups of tens to hundreds of thousands of stars held tightly together by gravity. Globular clusters have a definite, spherical shape and contain mostly reddish stars. The stars are closer together, closer to the center of the cluster. Globular clusters don't have much dust in them, the dust has alreadyformed into stars.

Spiral Galaxies

Galaxies are the biggest groups of stars and can contain anywhere from a few million stars to many billions of stars. Every star that is visible in the night sky is part of the Milky Way Galaxy. To the naked eye the closest major galaxy, the Andromeda Galaxy, looks like only a dim, fuzzy spot but that fuzzy spot contains one trillion stars.

Spiral galaxies spin, so they appear as a rotating disk of stars and dust, with a bulge in the middle, like the Sombrero Galaxy. Several arms spiral outward in the Pinwheel Galaxy and are appropriately called spiral arms. Spiral galaxies have lots of gas and dust and lots of young stars. Other galaxies are egg-shaped and called an elliptical galaxy. The smallest elliptical galaxies are as small as some globular clusters. Giant elliptical galaxies, on the other hand, can contain over a trillion stars. Elliptical galaxies are reddish to yellowish because they contain mostly old stars. Most elliptical galaxies contain very little gas and dust because they had already formed.

However, some elliptical galaxies contain lots of dust.

Figure 2.6: The Andromeda Galaxy. Image by NASA is in the public domain.

Irregular & Dwarf Galaxies

Galaxies that are not elliptical galaxies or spiral galaxies are irregular galaxies. Most irregular galaxies were once spiral or elliptical galaxies that were then deformed either by gravitational attraction to a larger galaxy or by a collision with another galaxy. Dwarf galaxies are small galaxies containing only a few million to a few billion stars. Dwarf galaxies are the most common type in the universe. However, because they are relatively small and dim, we don't see as many dwarf galaxies from Earth. Most dwarf galaxies are irregular in shape. However, there are also dwarf elliptical galaxies and dwarf spiral galaxies. Look back at the picture of the spiral galaxy, Andromeda. Next to our closest galaxy neighbor are two dwarf elliptical galaxies that are companions to the Andromeda Galaxy. One is a bright sphere to the left of the center, and the other is a long ellipse below and to the right of the center. Dwarf galaxies are often found near larger galaxies. They sometimes collide with and merge into their larger neighbors

Figure 2.7 An Irregularly Shaped Galaxy. Image by NASA is in the public domain.

THE MILKY WAY GALAXY

On a dark, clear night, you can see a milky band of light stretching across the sky. This band is the disk of a galaxy, the Milky Way Galaxy is our galaxy and is made of millions of stars along with a lot of gas and dust. Although it is difficult to know what the shape of the Milky Way Galaxy is because we are inside of it, astronomers have identified it as a typical spiral galaxy containing about 100 billion to 400 billion stars.

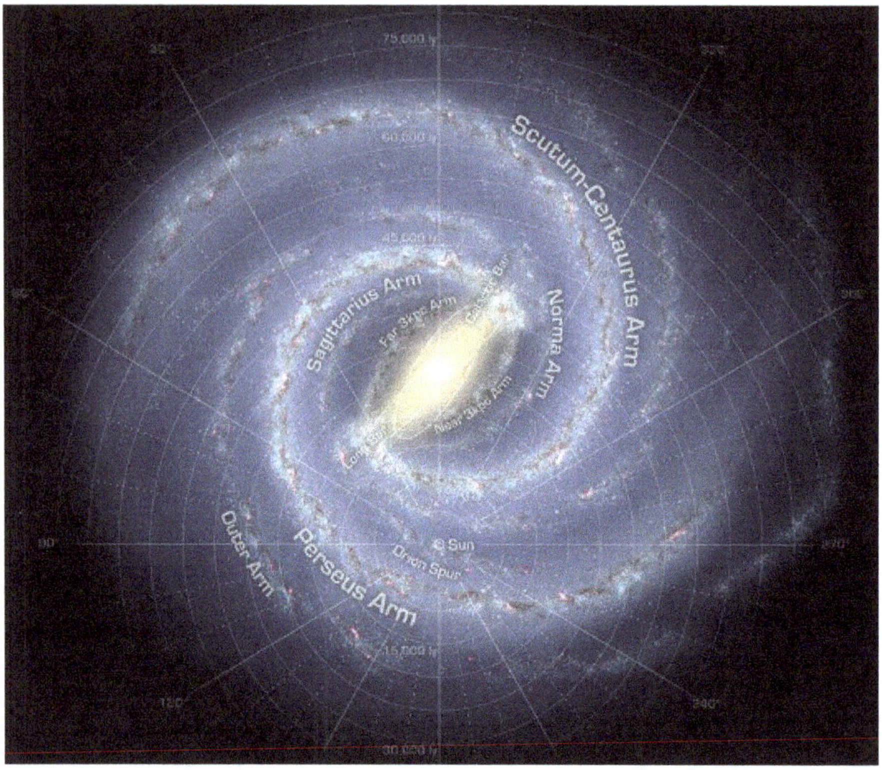

Figure 2.8 Artist's Conception of the Spiral Structure of the Milky Way. Image by NASA is in the public domain.

Artist's conception of the spiral structure of the Milky Way with two major stellar arms and a central bar. Using infrared images from NASA's Spitzer Space Telescope, scientists have discovered that the Milky Way's elegant spiral structure is dominated by just two arms wrapping off the ends of a central bar of stars. Previously, our galaxy was thought to possess four major arms.

Like other spiral galaxies, our galaxy has a disk, a central bulge, and spiral arms. The disk is about 100,000 light-years across and 3,000 light-years thick. Most of the Galaxy's gas, dust, young stars, and open clusters are in the disk. What data and evidence do astronomers find that lets them know that the Milky Way is a spiral galaxy?

- The shape of the galaxy as we see it.
- The velocities of stars and gas in the galaxy show a rotational motion.
- The gases, color, and dust are typical of spiral galaxies.

The central bulge is about 12,000 to 16,000 light-years wide and 6,000 to 10,000 light-years thick. The central bulge contains mostly older stars and globular clusters. Some recent evidence suggests the bulge might not be spherical but is instead shaped like a bar. The bar might be as long as 27,000 light-years long. The disk and bulge are surrounded by a faint, spherical halo, which also includes old stars and globular clusters. Astronomers have discovered that there is a gigantic black hole at the center of the galaxy.

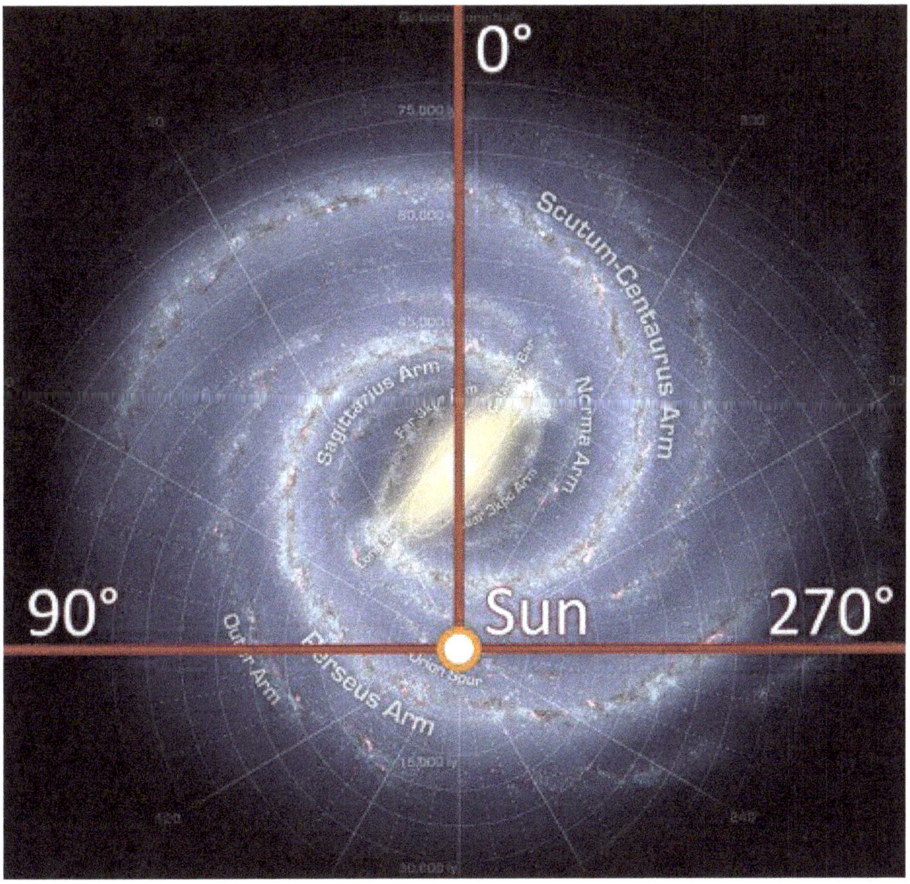

Figure 2.9 Grid Added to Annotated Milky Way. Image by NASA is in the public domain.

The Milky Way Galaxy is a significant place. Our solar system, including the Sun, Earth, and all the other planets, is within one of the spiral arms in the disk of the Milky Way Galaxy. Most of the stars we see in the sky are relatively nearby stars that are also in this spiral arm. Earth is about 26,000 light-years from the center of the galaxy, a little more than halfway out from the center of the galaxy to the edge. Just as Earth orbits the Sun, the Sun and solar system orbit the center of the Galaxy. One orbit of the solar system takes about 225 to 250 million years. The solar system has orbited 20 to 25 times since it formed 4.6 billion years ago. Astronomers have recently found that at the center of the Milky Way, and most other galaxies, is a supermassive black hole, though a black hole cannot be seen.

STAR ENERGY: NUCLEAR FUSION

A solar flare, also known as a corona, is a long filament of solar material, erupting out from the sun into space. The Sun is Earth's major source of energy, yet the planet only receives a small portion of its energy and the Sun is just an ordinary star. Many stars produce much more energy than the Sun. The energy source for all stars is nuclear fusion. Stars are made mostly of hydrogen and helium, which are packed so densely in a star that is the star's center the pressure is great enough to initiate nuclear fusion reactions. In a nuclear fusion reaction, the nuclei of two atoms combine to create a new atom. Most commonly, in the core of a star, two hydrogen atoms fuse to become a helium atom.

Figure 2.10 Solar Flare-- or Corona. Image by NASA is in the public domain.

Although nuclear fusion reactions require a lot of energy to get started, once they are going, they produce enormous amounts of energy. In a star, the energy from fusion reactions in the core pushes outward to balance the inward pull of gravity. This energy moves outward through the layers of the star until it finally reaches the star's outer surface. The outer layer of the star glows brightly, sending the energy out into space as electromagnetic radiation, including visible light, heat, ultraviolet light, and radio waves.

In particle accelerators, subatomic particles are propelled until they have attained almost the same amount of energy as found in the core of a star. When these particles collide head-on, new particles are created. This process stimulates the nuclear fusion that takes place in the cores of stars. The process also stimulates the conditions that allowed for the first helium atom to be produced from the collision of two hydrogen atoms in the first few minutes of the universe.

STAR CLASSIFICATION

Think about how the color of a piece of metal changes with temperature. A coil of an electric stove will start out black but with added heat will start to glow a dull red. With more heat, the coil turns a brighter red, then orange. At extremely high temperatures the coil will turn yellow- white, or even blue-white (it's hard to imagine the flame on your stove getting that hot). A star's color is also determined by the temperature of the star's surface. Relatively cool stars are red, warmer stars are orange or yellow, and extremely hot stars are blue or blue-white. Color is the most common way to classify stars. The table below shows the classification system. The class of a star is given by a letter. Each letter corresponds to a color, and also to a range of temperatures. Note that these letters don't match the color names; they are leftover from an older system that is no longer used. For most stars, the surface temperature is also related to size. Bigger stars produce more energy, so their surfaces are hotter. These stars tend toward bluish white. Smaller stars produce less energy. Their surfaces are less hot and so they tend to be yellowish.

Figure 2.11 Relative Sizes of the Planets in the Solar System and Several Well-Known Stars. Image by NASA is used under a Attribution-Share Alike 3.0 Unported license.

The Main Sequence

For most of a star's life, nuclear fusion in the core produces helium from hydrogen. A star in this stage is a main-sequence star. This term comes from the Hertzsprung-

Russell diagram shown here. For stars in the main sequence, the temperature is directly related to brightness. A star is on the main sequence as long as it can balance the inward force of gravity with the outward force of nuclear fusion in its core. The more massive a star, the more it must burn hydrogen fuel to prevent gravitational collapse. Because they burn more fuel, more massive stars have higher temperatures. Massive stars also run out of hydrogen sooner than smaller stars do. Our Sun has been a main-sequence star for about 5 billion years and will continue on the main sequence for about 5 billion more years. Very large stars may be in the main sequence for only 10 million years. Very small stars may last tens to hundreds of billions of years.

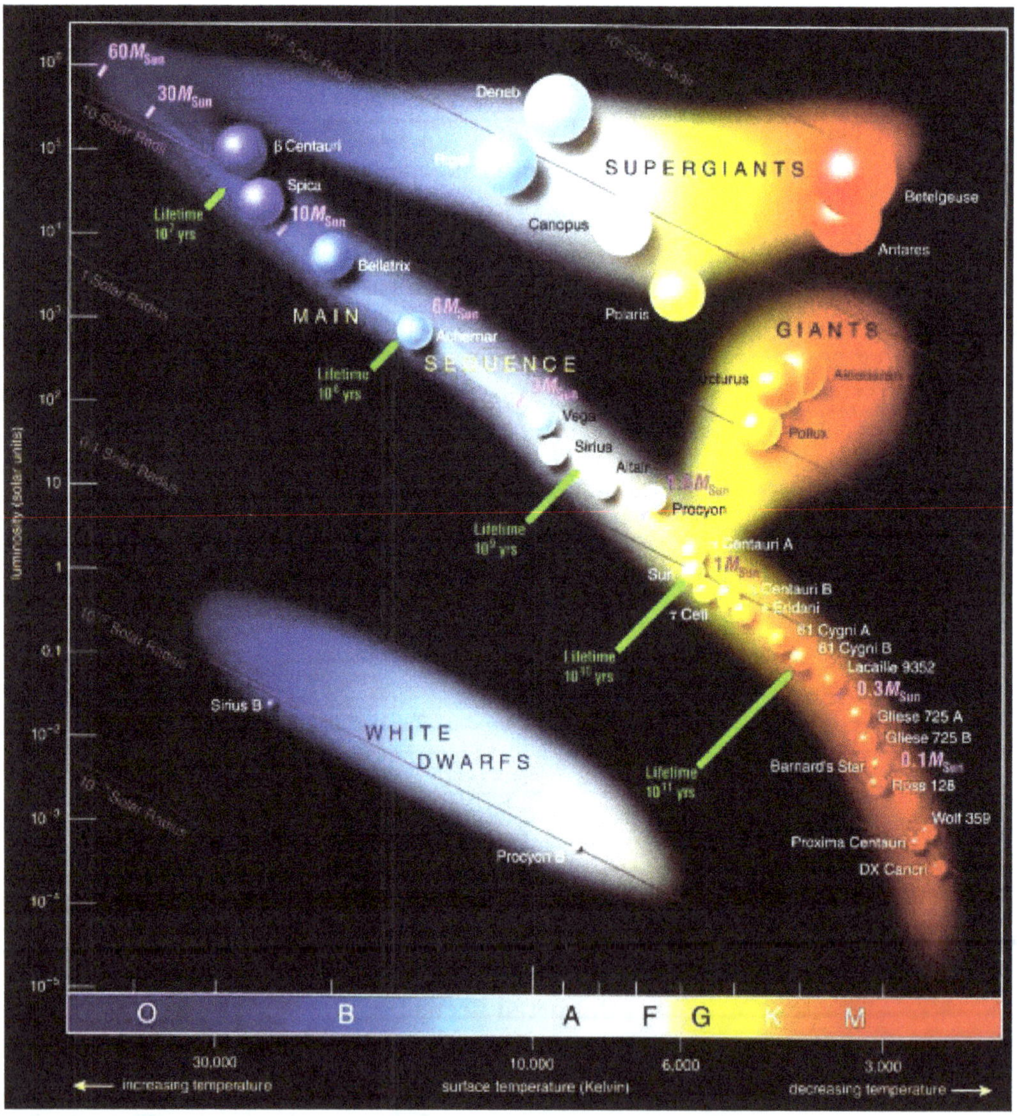

Figure 2.12 Main Sequence of a Stars Life. Image by NASA is in the public domain.

Red Giants & White Dwarfs

As a star begins to use up its hydrogen, it fuses helium atoms together into heavier atoms such as carbon. A blue giant star has exhausted its hydrogen fuel and is a transitional phase. When the light elements are mostly used up the star can no longer resist gravity and it starts to collapse inward. The outer layers of the star grow outward and cool. The larger, cooler star turns red and so is called a red giant. Eventually, a red giant burns up all of the helium in its core. What happens next depends on how massive the star is. A typical star, such as the Sun, stops fusion completely. Gravitational collapse shrinks the star's core to a white, glowing object about the size of Earth, called a white dwarf. A white dwarf will ultimately fade out.

Supergiants & Supernovas

A star that runs out of helium will end its life much more dramatically. When very massive stars leave the main sequence, they become red supergiants. Unlike a red giant, when all the helium in a red supergiant is gone, fusion continues. Lighter atoms fuse into heavier atoms up to iron atoms. Creating elements heavier than iron through fusion uses more energy than it produces so stars do not ordinarily form any heavier elements. When there are no more elements for the star to fuse, the core succumbs to gravity and collapses, creating a violent explosion called a supernova.

Figure 2.13 Image of the Crab Nebula. Image by NASA is in the public domain.

A supernova explosion contains so much energy that atoms can fuse together to produce heavier elements such as gold, silver, and uranium. A supernova can shine as brightly as an entire galaxy for a short time. All elements with an atomic number greater than that of lithium were created by nuclear fusion in stars.

Neutron Stars & Black Holes

After a supernova explosion, the leftover material in the core is extremely dense. If the core is less than about four times the mass of the Sun, the star becomes a neutron star. A neutron star is made almost entirely of neutrons, relatively large particles that have no electrical charge. If the core remaining after a supernova is more than about five times the mass of the Sun, the core collapses into a black hole. Black holes are so dense that not even light can escape their gravity. With no light, a black hole cannot be observed directly. But a black hole can be identified by the effect that it has on objects around it, and by radiation that leaks out around its edges.

Figure 2. 14: Simulated View of a Black Hole in Front of the Large Magellanic Cloud. Image by NASA is in the public domain.

SUMMARY

Astronomy is one of the oldest sciences, and early civilizations performed methodical observations of the night sky, and astronomical artifacts have been found from much earlier periods. However, the invention of the telescope was required before astronomy was able to develop into a modern science.

The Big Bang Theory is the most widely accepted cosmological explanation of how the universe formed. At its simplest, it says the universe as we know it started with a small singularity, then inflated over the next 13.8 billion years to the cosmos that we know today Image result for milky way summary.

The Milky Way contains over 200 billion stars, and enough dust and gas to make billions more. The solar system lies about 30,000 light-years from the galactic center and about 20 light-years above the plane of the galaxy. More than half the stars found in the Milky Way are older than the 4.5-billion-year-old sun.

3 Chapter | INTRODUCTION TO GEOLOGY & GEOLOGIC TIME

Figure 3.1 Student at Ubehebe Crater in Death Valley, California. Image by Jeremy Patrich is used under a CC-BY 4.0 license.

GOALS & OBJECTIVES OF THIS UNIT

- Explain what geology is, how it incorporates the other sciences, and how it is differentfrom the other sciences.
- Discuss why we study Earth and what type of work geologists do.
- Use the notation for geological time, gain an appreciation for the vastness of geologicaltime, and describe how very slow geological processes can have enormous impacts over time. Apply basic geological principles to the determination of the relative ages of rocks.
- Explain the difference between relative and absolute age-dating techniques.
- Summarize the history of the geological time scale and the relationships between eons,eras, periods, and epochs.

WHAT IS GEOLOGY?

In its broadest sense, geology is the study of Earth, it's interior and its exterior surface, the rocks and other materials that are around us, the processes that have resulted in the formation of those materials, the water that flows over the surface, and lies underground, the changes that have taken place over the vastness of geologic time, and the changes that we can anticipate will take place in the near future. Geology is a science, meaning that we use deductive reasoning and scientific methods to understand geological problems. It is, arguably, the most integrated of all of the sciences because it involves the understanding and application of all of the other sciences: physics, chemistry, biology, mathematics, astronomy, and others.

But unlike most of the other sciences, geology has an extra dimension, that of time, deep time, billions of years of it. Geologists study the evidence that they see around them, but in most cases, they are observing the results of processes that happened thousands, millions, and even billions of years in the past. Those were processes that took place at incredibly slow rates, millimeters per year to centimeters per year, but because of the amount of time available, they produced massive results.

The Sierra Nevada is a mountain range in the Western United States, between the Central Valley of California and the Great Basin. The vast majority of the range lies in the state of California, although the Carson Range spur lies primarily in Nevada. The Sierra Nevada is part of the American Cordillera, a chain of mountain ranges that consists of an almost continuous sequence of such ranges that form the western "backbone" of North America, Central America, South America, and Antarctica.

The Sierra runs 400 miles (640 km) north-to-south and is approximately 70 miles (110 km) across east-to-west. Notable Sierra features include The General Sherman, the largest tree in the world by volume; Lake Tahoe, the largest alpine lake in North America; Mount Whitney at 14,508 ft (4,422 m), the highest point in the contiguous United States; and Yosemite Valley sculpted by glaciers from one-hundred-million-year-old granite, containing high waterfalls. The Sierra is home to three national parks, twenty wilderness areas, and two national monuments. These areas include Yosemite, Sequoia, and Kings Canyon National Parks; and Devils Postpile National Monument.

Geology is also about understanding the evolution of life on Earth; about discovering resources such as metals and energy; about recognizing and minimizing the environmental implications of our use of those resources; and about learning how to mitigate the hazards related to earthquakes, volcanic eruptions, and slope failures.

What Do Geologists Do?

Geologists are involved in a range of widely varying occupations with one thing in common: the privilege of studying this fascinating planet. Many geologists work in the resource industries, including mineral exploration and mining and energy exploration and extraction. Other major areas where geologists work include hazard assessment and mitigation (e.g., assessment of risks from slope failures, earthquakes, and volcanic eruptions); water supply planning, development, and management; waste management; and assessment of geological issues on construction projects such as highways, tunnels, and bridges. Most geologists are employed in the private sector, but many work for government-funded geological organizations, such as the United States Geological Survey, (USGS).

Figure 3.2 Faculty Explaining Physical Weathering of Rock Material. Image by Jeremy Patrich is used under a CC-BY 4.0 license.

Many people are attracted to geology because they like to be outdoors, and many geological opportunities involve fieldwork in places that are as amazing to see as they are interesting to study. But a lot of geological work is also completed back in offices or laboratories.

GEOLOGICAL TIME

The Geologic Time Scale and the basic outline of Earth history were worked out long before we had any scientific means of assigning numerical units of age like years to events of Earth history. Working out Earth history depended on realizing some key principles of relative time. Nicholas Steno introduced a basic understanding of stratigraphy (the study of layered rocks) in 1669 with the basic principles of stratigraphy. William Smith (1769-1839), working with the strata of the English coal mines, noticed that strata and their sequence were consistent throughout the region and eventually produced the first national geologic map of Britain becoming known as "the Father of English Geology." Using Steno's principles, a relative time scale was developed in the nineteenth century with names derived from areas studied and characteristics of the rocks in those areas. The figure below shows the names applied to units and subunits of the Geologic Time Scale. Using this time scale as a calendar, all events of Earth history can be placed in order without ever knowing the numerical age.

A useful mechanism for understanding geological time is to scale it all down into one year. The origin of the solar system and Earth at 4.57 Ga (billion years ago) would be represented by January 1, and the present year would be represented by the last tiny fraction of a second on New Year's Eve. At this scale, each day of the year represents 12.5 million years; each hour represents about 500,000 years; each minute represents 8,694 years, and each second represents 145 years. Some significant events in Earth's history, as expressed on this time scale, are summarized in the table below.

Table 3.1 Geological Events

Geologic Event	Approximate Date	Calendar Equivalent
Formation of oceans and continents	4.5-4.4 Ga	January 1st
Evolution of the first primitive life forms	3.8 Ga	Early March
Stromatolites & the origin of Earth's Oxygen	3.5 Ga	April 1st
Ediacaran Fauna	600 Ma	November 11th
Cambrian Explosion	545 Ma	November 16th
Animals first crawled onto land	360 Ma	December 1st
Extinction of the non-avian dinosaurs	65 Ma	December 26th
Beginning of the Pleistocene ice age	1 Ma or 200 ka	8:00 pm on December 31st
Native Americans made the Channel Islands their home	10 ka	11:59 pm on December 31st
The arrival of the first Europeans on the northern west coast of North America	250 years ago	Two seconds before mid night on December 31st

The Geological Time Scale

William "Strata" Smith worked as a surveyor in the coal-mining and canal-building industries in southwestern England in the late 1700s and early 1800s. While doing his work, he had many opportunities to look at the Paleozoic and Mesozoic sedimentary rocks of the region, and he did so in a way that few had done before. Smith noticed the textural similarities and differences between rocks in different locations, and more importantly, he discovered that fossils could be used to correlate rocks of the same age. Smith is credited with formulating the principle of faunal succession (the concept that specific types of organisms lived during different time intervals), and he used it to great effect in his monumental project to create a geological map of England and Wales, published in 1815.

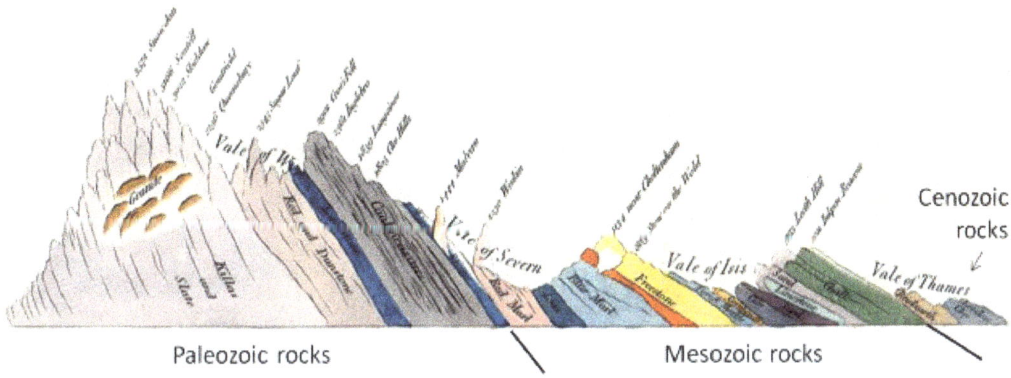

Figure 3.3 William Smith's "Sketch of the Succession of Strata and Their Relative Altitudes. Image by NASA is in the public domain.

Inset into Smith's great geological map is a small diagram showing a schematic geological cross- section extending from the Thames estuary of eastern England the west coast of Wales. Smith shows the sequence of rocks, from the Paleozoic rocks of Wales and western England, through the Mesozoic rocks of central England, to the Cenozoic rocks of the area around London.

Although Smith did not put any dates on these, because he didn't know them, he was aware of the principle of superposition (the idea, developed much earlier by the Danish theologian and scientist Nicholas Steno, that young sedimentary rocks form on top of older ones), and so he knew that this diagram represented a stratigraphic column. And because almost every period of the Phanerozoic is represented along that section through Wales and England, it is a primitive geological time scale.

Smith's work set the stage for the naming and ordering of the geological periods, which wasinitiated around 1820, first by British geologists, and later by other European geologists. Manyof the periods are named for places where rocks of that age are found in Europe, such as Cambrian for Cambria (Wales), Devonian for Devon in England, Jurassic for the Jura Mountains in France and Switzerland, and Permian for the Perm region of Russia. Some are named for the type of rock that is common during that age, such as Carboniferous for the coal- and carbonate-bearing rocks of England, and Cretaceous for the chalks of England and France.

The early time scales were only relative because 19th-century geologists did not know the ages of the rocks. That information was not available until the development of isotopic dating techniques early in the 20th century.

The geological time scale is currently maintained by the International Commission on Stratigraphy (ICS), which is part of the International Union of Geological Sciences. The time scale is continuously being updated as we learn more about the timing and nature of past geological events.

Geological time has been divided into four eons: Hadean, Archean, Proterozoic, and Phanerozoic, and as shown below, the first three of these represent almost 90% of Earth's history. The last one, the Phanerozoic (meaning "visible life"), is the time that we are most familiar with because Phanerozoic rocks are the most common on Earth, and they contain evidence of life forms that we are all somewhat familiar with.

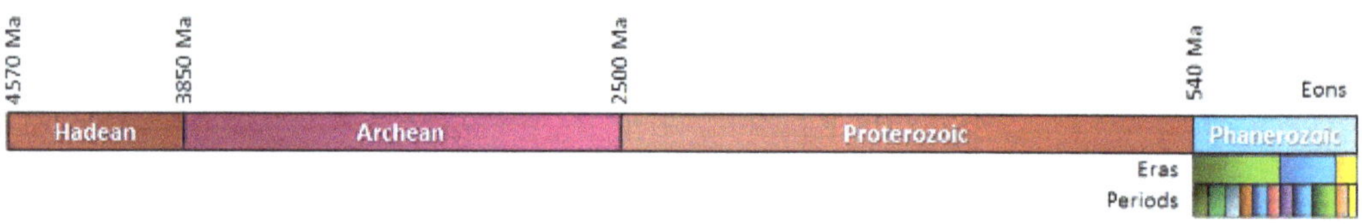

Figure 3.4 The Eons of Earth's History. Image is used under a CC-BY 4.0 license.

The Phanerozoic, the past 540 Ma of Earth's history, is divided into three eras: the Paleozoic ("early life"), the Mesozoic ("middle life"), and the Cenozoic ("new life"), and each of these is divided into several periods. Most of the organisms that we share Earth with evolved at various times during the Phanerozoic.

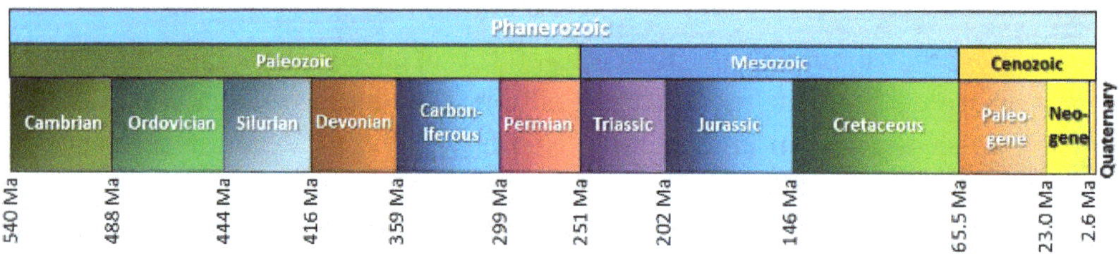

Figure 3.5 The Eras and Periods of the Phanerozoic. Image is used under a CC-BY 4.0 license.

The Cenozoic, which represents the past 65.5 Ma, is divided into three periods: Paleogene, Neogene, and Quaternary, and seven epochs. Dinosaurs became extinct at the start of the Cenozoic, after which birds and mammals radiated to fill the available habitats. Earth was very warm during the early Eocene and has steadily cooled ever since. Glaciers first appeared on Antarctica in the Oligocene and then on Greenland in the Miocene and covered much of North America and Europe by the Pleistocene. The most recent of the Pleistocene glaciations ended around 11,700 years ago. The current epoch is known as the Holocene. Epochs are further divided into ages or stages.

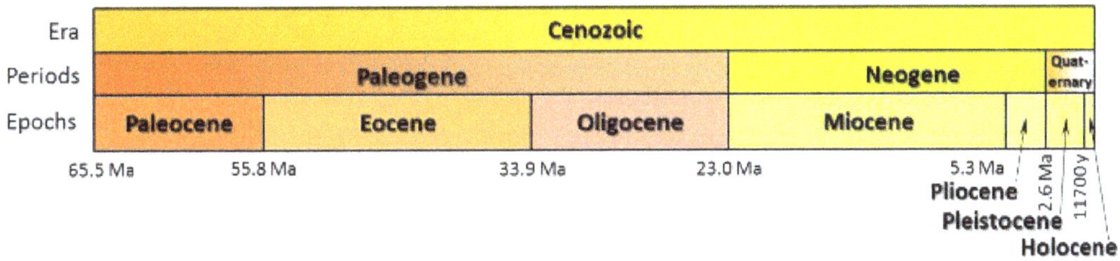

Figure 3.6 The Periods and Epochs of the Cenozoic. Image is used under a CC-BY 4.0 license.

Most of the boundaries between the periods and epochs of the geological time scale have been fixed based on significant changes in the fossil record. For example, as already noted, the boundary between the Cretaceous and the Paleogene coincides exactly with the extinction of the dinosaurs. That's not a coincidence. Many other types of organisms went extinct at this time, and the boundary between the two periods marks the division between sedimentary rocks with Cretaceous organisms below and Paleogene organisms above.

Figure 3.7 Geological Time Spiral. Image is in the public domain.

GEOLOGIC DATING

Geological Dating is a technique used in Geology to date a certain type of rock that contains radiometric elements and those radiometric elements decay at a constant rate. This unit will discuss several different types of dating, both relative and absolute.

Relative Dating

Relative dating is the process of determining if one rock or geologic event is older than or younger than another, without knowing the specific age (e.g., number of years ago the object was formed). The principles of relative time are simple, even obvious now, but were not generally accepted by scholars until the Scientific Revolution of the 17th and 18th centuries.

James Hutton realized that geologic processes are slow and his ideas on uniformitarianism (e.g., "the present is the key to the past") provided a basis for interpreting the rocks of the Earth in terms of scientific principles.

Stratigraphy is the study of layered sedimentary rocks. Below are a few principles of relative time that are used in all of geology, but especially useful in stratigraphy.

- Principle of Superposition: In an otherwise undisturbed sequence of sedimentary strata (rock layers), the layers on the bottom are the oldest and the layers above are younger.
- Principle of Original Horizontality: Layers of rocks deposited from above in a gravity
- field, such as sediments and lava flows, originally were laid down horizontally. This holds true except for the margins of basins, where the strata can slope slightly downward into the basin.
- Principle of Lateral Continuity: Within the depositional basin in which they form, strata are continuous in all directions until they thin out at the edge of that basin. Of course, all strata eventually end, either by hitting a geographic barrier or by a depositional process being too far from its source, either a sediment source or a volcano. Strata that are subsequently cut by a canyon remain continuous on either side of the canyon.
- Principle of Fossil Succession: Assemblages of fossils contained in strata are unique to the time they lived and can be used to correlate rocks of the same age across a wide geographic distribution. Evolution has produced a succession of life whose fossils are unique to the units of the Geologic time Scale.

Figure 3.8 Example of Superposition, Red Rock Canyon, California. Image by Jeremy Patrich is used under a CC-BY 4.0 license.

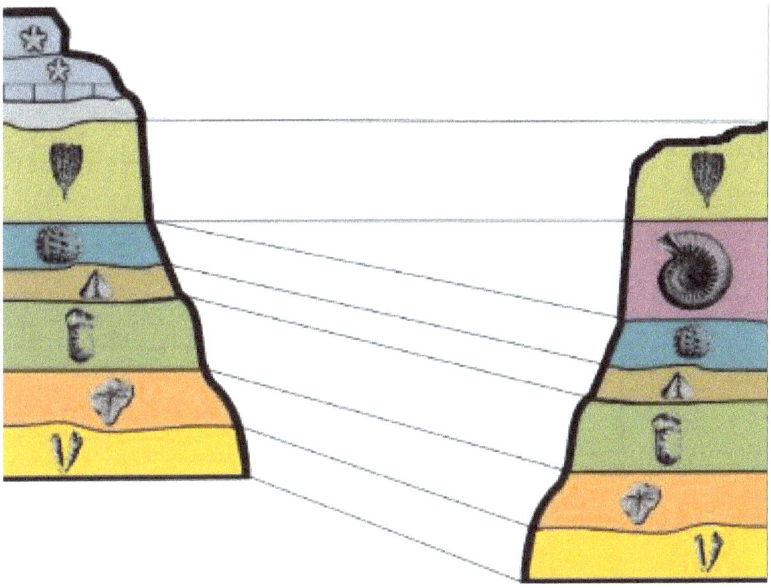

Figure 3.9 Fossil Succession Showing Correlation Among Strata. Image is used under a CC-BY 4.0 license.

Absolute Dating

Relative time allows science to tell the story of the Earth, but does not provide specific numeric ages of events, and thus, the rate at which geologic processes operate. Based on Hutton's Uniformity Principle (Uniformitarianism), early geologists surmised that geological processes work slowly and that the Earth is very old. Because science advances as the technology of its tool's advances, the discovery of radioactivity in the late 1800s provided a new scientific tool by which actual ages in years can be assigned to mineral grains within a rock. Later we will identify how Earth history is understood using relative dating principles without actually knowing the numerical age of events. This was how scientists of that time interpreted Earth history, until the end of the 19th-century when radioactivity was discovered. This discovery introduced a new dating technology that allows scientists to determine specific numeric ages of some rocks, called absolute dating. The next sections discuss this absolute dating system called radio- isotopic dating.

Radio-Isotopic Dating

Given a sample of rock, how is the dating procedure carried out? Using chemical analysis, the parent elements and daughter products can be separated from the mineral. Remember that elements behave chemically due to their atomic number. In the case of uranium, both the 238U and 235U isotopes are chemically separated out together, as are the 206Pb and 207Pb. An instrument called a mass spectrometer then separates the uranium isotopes from each other as well as the lead isotopes from each other by passing beams of the isotopes through a magnetic field. As these isotopic beams pass through the instrument, the path of the heavier isotope is deflected less so the two beams strike a sensor at different places. From the intensity of each beam, the amount of parent and daughter products is determined, and from this ratio, the age can be calculated.

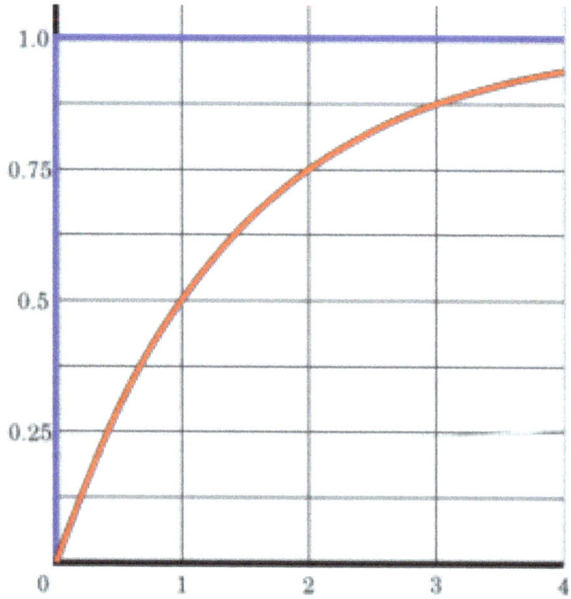

Figure 3.10 Graph of the Number of Half-Lives Image is used under a CC-BY 4.0 license

Here is a simple example of age calculation using the ratio of daughter product to parent isotope. When the mineral initially forms, there is a 100% parent isotope and 0% daughter and the ratio of daughter to parent (D/P) is 0. After one half-life, half the parent has decayed so there is 50% parent and 50% daughter. The ratio is then 1. After two half-lives, there is 25% parent and 75% daughter, and the ratio is 3. This can be further calculated for a series of half- lives as shown in the table below. Note that after about ten half-lives, the amount of parent remaining is so small that accurate chemical analysis of the parent is difficult, and the accuracy of the method is diminished. Ten half-lives are generally considered the upper limit for use of an isotope for radio-isotopic dating. Modern applications of this method have achieved remarkable accuracy of plus or minus two million years in 2.5 billion years (that's ±0.055%).

Considering the uranium/lead technique, in any given sample analysis, there are two separate clocks running at the same time, 238U and 235U. The existence of these two clocks in the same sample gives a cross-check on each other. Many geological samples contain multiple parent/daughter pairs so cross-checking clocks show radio-isotopic dating to be highly reliable.

Carbon Dating

Another radio-isotopic dating method involves carbon and is useful for dating archaeologically important samples containing organic substances like wood or bone. Carbon dating uses the unstable isotope carbon-14 (14C) and the stable isotope carbon-12 (12C). Carbon-14 is constantly being created in the atmosphere by the interaction of cosmic particles with atmospheric nitrogen-14 (14N). The cosmic particles include neutrons that strike the nitrogen nucleus kicking out a proton but leaving the neutron in the nucleus. The atomic number is reduced by one from 7 to 6 forming carbon and the mass number remains the same at 14. The 14C quickly bonds with oxygen in the atmosphere to form carbon dioxide which mixes with the other atmospheric carbon dioxide and is incorporated into living matter. Thus, while an organism is alive the ratio of 14C/12C in its body doesn't change since it is constantly exchanging with the atmosphere. However, when it dies, the radiocarbon clock starts ticking as the 14C decays back to 14N by beta decay with a half-life of 5,730 years. The radiocarbon dating technique is thus useful for about ten half-lives back 57,300 years or so.

Since radio-isotopic dating relies on parent and daughter ratios and the amount of parent 14C needs to be known, early applications of 14C dating assumed the production and concentration of 14C in the atmosphere for the last 50,000 years or so was the same as today. But the production of CO2 since the Industrial Revolution by combustion of fossil fuels (in which 14C long ago decayed) has diluted 14C in the atmosphere leading to potential errors in this assumption. Other factors affecting the estimates of the composition of parent carbon in the atmosphere have also been studied. Comparisons of carbon ages with tree ring data and other data for known events have allowed calibration for the reliability of the radiocarbon method which is primarily used in archaeology and very recent geologic events. Taking into account these factors, carbon-14 dating is a reliable dating method in this range.

SUMMARY

All the events of earth history can be placed in sequence using the principles of relative time called the five Principles of Stratigraphy. The Geologic Time Scale was completely worked out in the 19th Century using these principles and used as a calendar for telling the story of the earth.

Hutton's Uniformity Principle, knowing how natural processes work in the present is key to understanding the past, provided a means to study and understand the processes involved in the events that have shaped the earth to its present form.

The discovery of radioactivity in the late 1800s provided a tool to measure the actual ages of the events of earth history. However, certain types of rocks and minerals are better suited for dating and certain assumptions about those rocks and minerals require care and precaution in interpreting ages.

Geologic time is now known to be vast and to have provided plenty of time for the evolution of the planet and life upon it to have taken place to produce the earth as we see it and all the life forms on it.

4 Chapter | MAPPING EARTH'S SURFACE

Figure 4.1 Lowest Elevation in California, Badwater Basin -282ft in Death Valley, California.

GOALS & OBJECTIVES OF THIS UNIT

- Understand a map's scale, projections, and ways of telling the map user what the map ismeasuring on Earth's surface.
- Explore the concepts of scale, resolution, and projection.
- Identify contour intervals on basic topographic maps.
- Allow students to interpret the distortions of each projection, and to explain how the point of tangency creates different styles of maps and presents different information.

- Provide students an understanding of latitude and longitude, and how to use this andother coordinate systems.

WHAT IS CARTOGRAPHY?

Cartography is the study, practice, and interpretation of maps. Combining science, aesthetics, and technique, cartography builds on the premise that reality can be modeled in ways that communicate spatial information effectively.

The fundamental problems of traditional cartography are to:

- Set the map's agenda and select the traits of the object to be mapped. This is the concern of map editing. Traits may be physical, such as roads or land masses, or maybe abstract, such as typonyms or political boundaries.
- Represent the terrain of the mapped object on flat media. This is the concern of mapprojections.
- Reduce the complexity of the characteristics that will be mapped. This is also the concern of generalization.
- Orchestrate the elements of the map to best convey its message to its audience. This isthe concern of design.

Map Scale

The Earth's surface has an area of over 500 million km2 and any picture of the earth that you can easily carry can only show general outlines of continents and countries. When we visually represent a region of the world on a map, we must reduce its size to fit within the boundaries of the map. Map scale measures how much the features of the world are reduced to fit on a map; or more precisely, map scale shows the proportion of a given distance on a map to the corresponding distance on the ground in the real world. The map scale is represented by a representative fraction, graphic scale, or verbal description.

Representative Fraction

The most commonly used measure of map scale is the representative fraction (RF), where a map scale is shown as a ratio. With the numerator always set to 1, the denominator represents how much greater the distance is in the world. The figure below shows a topographic map with an RF of 1:24,000, which means that one unit on the map represents 24,000 units on the ground. The representative fraction is accurate regardless of which units are used; the RF can be measured as 1 centimeter to 24,000 centimeters, one inch to 24,000 inches, or any other unit.

Graphic Scale

Scale bars are graphical representations of distance on a map. The figure has scale bars for 1 mile, 7000 feet, and 1 kilometer. One important advantage of graphic scales is that they remain true when maps are shrunk or magnified.

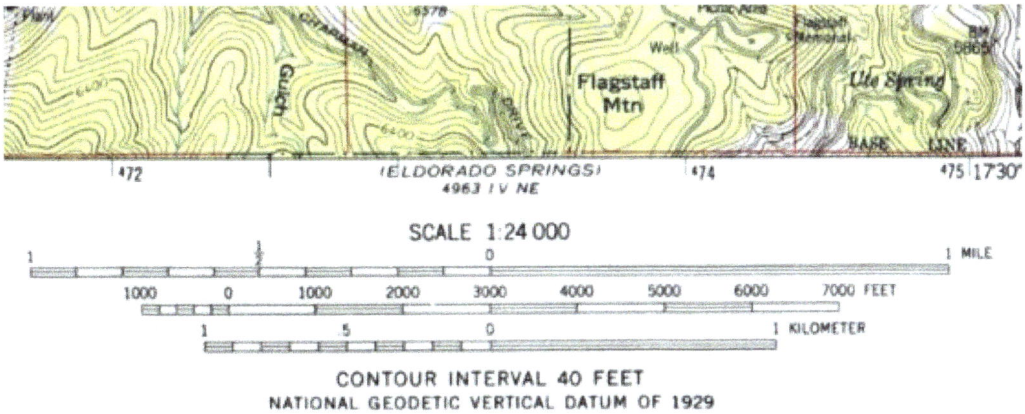

Figure 4.2 Representative Fraction & Scale Bars from a USGS Topographic Map. Image is used under a Creative Commons by-nc-sa 3.0 license.

Verbal Description

Some maps, especially older ones, use a verbal description of scale. For example, it is common to see "one inch represents one kilometer" or something similar written on a map to give map users an idea of the scale of the map.

Mapmakers use a scale to describe maps as being small-scale or large-scale. This description of the map scale as large or small can seem counterintuitive at first. A 3-meter by 5-meter map of the United States has a small map scale while a college campus map of the same size is large- scale. Scale descriptions using the RF provide one way of considering a scale, since 1:1000 is larger than 1:1,000,000. Put differently, if we were to change the scale of the map with an RF of 1:100,000 so that a section of road was reduced from one unit to, say, 0.1 units in length, we would have created a smaller-scale map whose representative fraction is 1:1,000,000. In general, the larger the map scale, the more detail that is shown.

CONTOUR LINES

Contour lines are the greatest distinguishing feature of a topographic map. Contour lines are lines drawn on a map connecting points of equal elevation, meaning if you physically followed a contour line, the elevation would remain constant. Contour lines show elevation and the shape of the terrain. They're useful because they illustrate the shape of the land surface (topography) on the map. Here's an easy way to understand how to interpret contour lines: Take an objectlike a ball or a pile of laundry and shine a red laser pointer along the object's side. The line you see will look like a contour line on a topographic map.

Topographic maps show lines for certain elevations only. These lines are evenly spaced apart. We call this spacing the contour interval. For example, if your map uses a 10-foot contour interval, you will see contour lines for every 10 feet (3 meters) of elevation lines at 0, 10, 20, 30, 40, and so on. Different maps use different intervals, depending on the topography. If, for example, the general terrain is quite elevated, the map might run at 80- to even 100-foot (24.4- to 30.5-meter) intervals. This makes it easier to read the map, as too many contour lines would be difficult to work with.

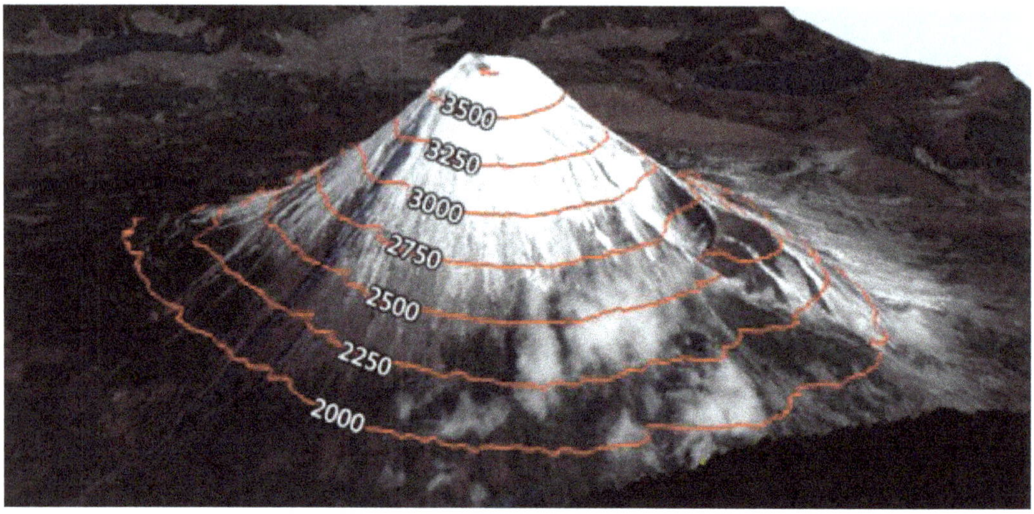

Figure 4.3 Mount Fuji with Contour Lines (USGS). Image used with permission.

To make topographic maps easier to read, every fifth contour line is an index contour. Because it's impractical to mark the elevation of every contour line on the map, the index contour lines are the only ones labeled. The index contours are a darker or wider brown line in comparison to the regular contour lines. You'll see the elevations marked on the index contour lines only. To determine elevations, pay attention to the amount of space between lines. If the contours are close together, you're looking at a steep slope. If the contours have wide spaces in between or aren't there at all, the terrain is relatively flat.

EXTENT VERSUS RESOLUTION

The extent of a map describes the area visible on the map, while resolution describes the smallest unit that is mapped. You can think of the extent as describing the region to which the map is zoomed. The extent of the map below is national as it encompasses the contiguous.

United States, while the resolution is the state because states are the finest level of spatial detail that we can see.

We often choose mapping resolutions intentionally to make the map easier to understand. For example, if we tried to display a map with a national extent at the resolution of census blocks, the level of detail would be so fine, and the boundaries would be so small that it would be difficult to understand anything about the map.

Balancing extent and resolution are often one of the most important and difficult decisions a cartographer must make. The figure below offers two more examples of the difference between extent and resolution.

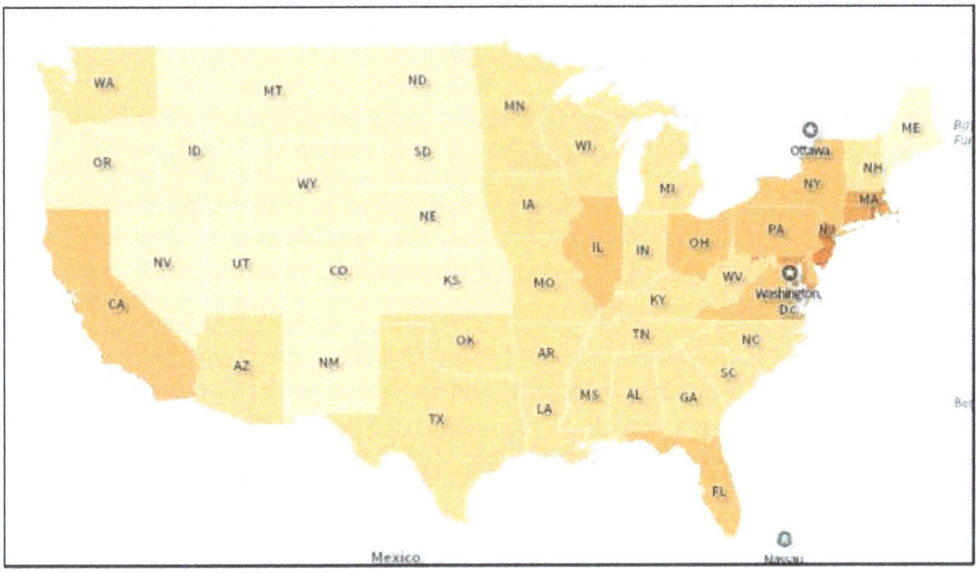

Figure 4.4 Map Showing Population Density Over National Extent & State Resolution, Data from US Census. Image by Steve Manson is licensed under a CC BY-NC-SA 4.0 license.

Coordinated & Projections

Locations on the Earth's surface are measured in terms of coordinates, a set of two or more numbers that specify a location to some reference system. The simplest system of this kind is a Cartesian coordinate system, named for the 17th -century mathematician and philosopher René Descartes.

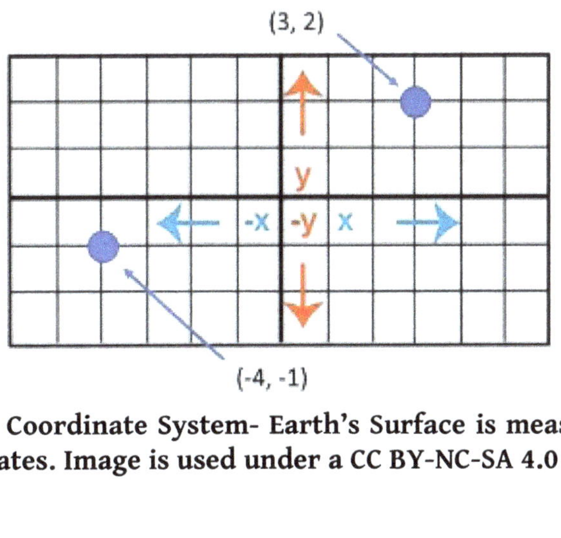

Figure 4.5 Sample Coordinate System- Earth's Surface is measured in Terms of Coordinates. Image is used under a CC BY-NC-SA 4.0 license.

A Cartesian coordinate system, like the one below, is simply a grid formed by putting together two measurement scales, one horizontal (x) and one vertical (y). The point at which both x and y equal zero is called the origin of the coordinate system. In the figure, the origin (0,0) is located at the center of the grid (the intersection of the two bold lines). All other positions are specified relative to the origin, as seen with the points at (3, 2) and (-4, -1).

The geographic coordinate system is designed specifically to define positions on the Earth's roughly spherical surface. Instead of the two linear measurement scales x and y, as with a Cartesian grid, the geographic coordinate system uses an east-west scale, called longitude that ranges from +180° to -180°. Because the Earth is round, +180° (or 180° E) and -180° (or 180° W) are the same grid line, termed the International Date Line. Opposite the International Date Line is the prime meridian, the line of longitude defined as 0°. The north-south scale, called latitude, ranges from +90° (or 90° N) at the North Pole to -90° (or 90° S) at the South Pole. In simple terms, longitude specifies positions east and west and latitude specify positions north and south. At higher latitudes, the length of parallels decreases to zero at 90° North and South.

Lines of longitude are not parallel but converge toward the poles. Thus, while a degree of longitude at the equator is equal to a distance of about 111 kilometers, or about 69 mi, that distance decreases to zero at the poles.

Projection is the process of making a two-dimensional map from a three-dimensional globe. We can think of the earth as a sphere. In reality, it is more of an ellipsoid with a few bulges, but it is fine to think of it as a sphere. To get a sense of how difficult this process can be, imagine peeling the skin from an orange and trying to lay the skin flat.

Figure 4.6 Flattened Orange Peel Representing Earth as a Flat Surface. Image is used under a CC BY-NC-SA 3.0.

As you peel and flatten the skin, you will encounter several problems:

- Shearing – stretching the skin in one or more directions.

- Tearing – causing the skin to separate.
- Compressing – forcing the skin to bunch up and condense.
- Cartographers face the same three issues when they try to transform the three-dimensional globe into a two-dimensional map. If you had a globe made of paper, you could carefully try to 'peel' it into a flat piece of paper, but you would have a big mess on your hands. Instead, cartographers use projections to create useable two-dimensional maps.

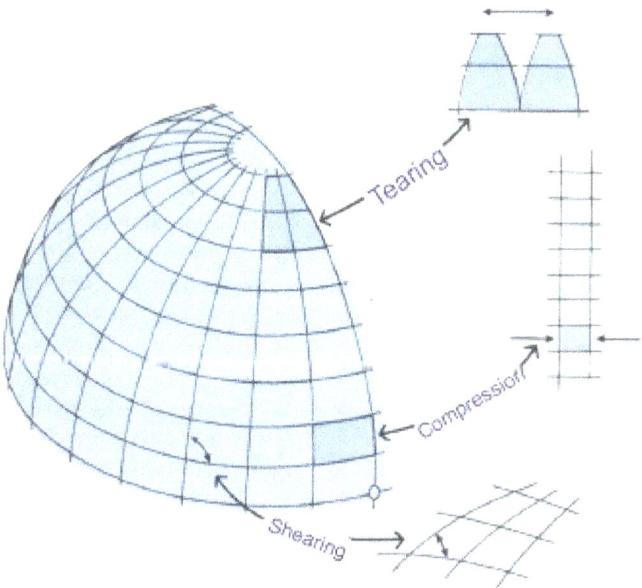

Figure 4.7 Shearing, Compression & Tension Distortion of a Globe (Steve Manson). Image is used under a CC BY-NC-SA 4.0 license.

PROJECTION MECHANICS

The term "map projection" refers to both the process and product of transforming spatial coordinates on a three-dimensional sphere to a two-dimensional plane. In terms of actual mechanics, most projections use mathematical functions that take as inputs locations on the sphere and translate them into locations on a two-dimensional surface.

It is helpful to think about projections in physical terms. If you had a clear globe the size of a beach ball and placed a light inside this globe, it would cast shadows onto a surrounding surface. If this surface were a piece of paper that you wrapped around the globe, you couldcarefully trace these shadows onto the paper, then flatten out this piece of paper and have your projection!

Most projections transform part of the globe to one of three "developable" surfaces, so-called because they are flat or can be made flat: plane, cone, and cylinder. The resulting projections are called planar, conical, and cylindrical. We use developable surfaces because they eliminate tearing, although they will produce shearing and compression. Of these three problems, tearing is seen as the worst because you would be making maps with all sorts of holes in them. As we see below, however, there are times when you can create maps with tearing, and they are quite useful.

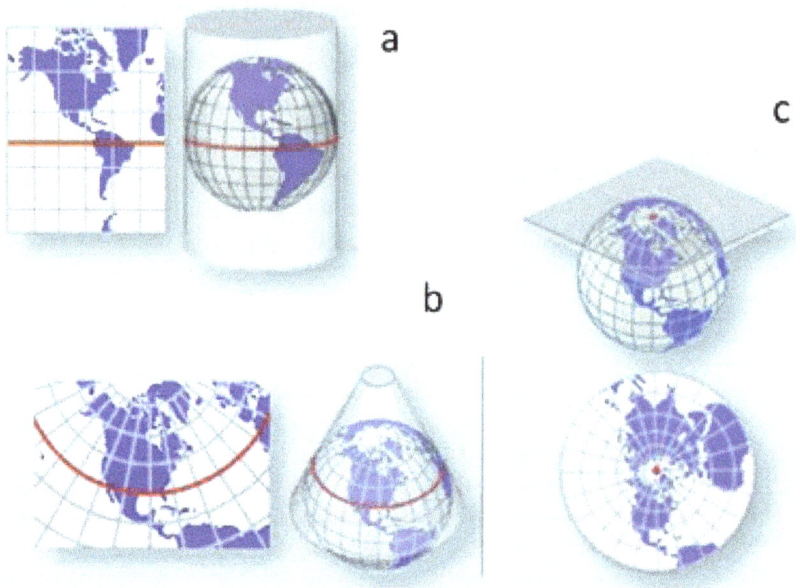

Figure 4.8 Red Lines or Dots Mark the Tangent Line or Point Respectively. The Flat Surface Touches the Globe & it is The Point on the Projected Map Which Has the Least Distortion. Image is in the public domain.

The place where the developable surface touches the globe is known as the tangent point or tangent line. Maps will most accurately represent objects on the globe at these tangent points or lines, with distortion increasing as you move farther away due to shearing and compression. It is for this reason that cylinders are often used for areas near the equator (great circles), cones used to map the mid-latitudes (small circles), and planes used for Polar Regions (points).

For beginning mapmakers, understanding the exact mechanics of projections doesn't matter asmuch as knowing which map properties are maintained or lost with the choice of projection.

Projections must distort features on the surface of the globe during the process of making them flat because projection involves shearing, tearing, and compression. Since no projection can preserve all properties, it is up to the mapmaker to know which properties are most importantfor their purpose and to choose an appropriate projection. The properties we will focus on are shape, area, and distance.

Note that distortion is not necessarily tied to the type of developable surface but rather to the way the transformation is done with that surface. It is possible to preserve any one of the three properties using any of the developable surfaces. One way of looking at the problem is with distortion ellipses. These help us to visualize what type of distortion a map projection has caused, how much distortion has occurred, and where it has occurred. The ellipses show how imaginary circles on the globe are deformed as a result of a particular projection. If no distortion had occurred in projecting a map, all of the ellipses would be the same size and circular.

Conformal

Conformal projections preserve shape and angle, but strongly distorts area in the process. For example, with the Mercator projection, the shapes of coastlines are accurate on all parts of the map, but countries near the poles appear much larger relative to countries near the equator than they are. For example, Greenland is only 7% of the land area of Africa, but it appears to be just as large.

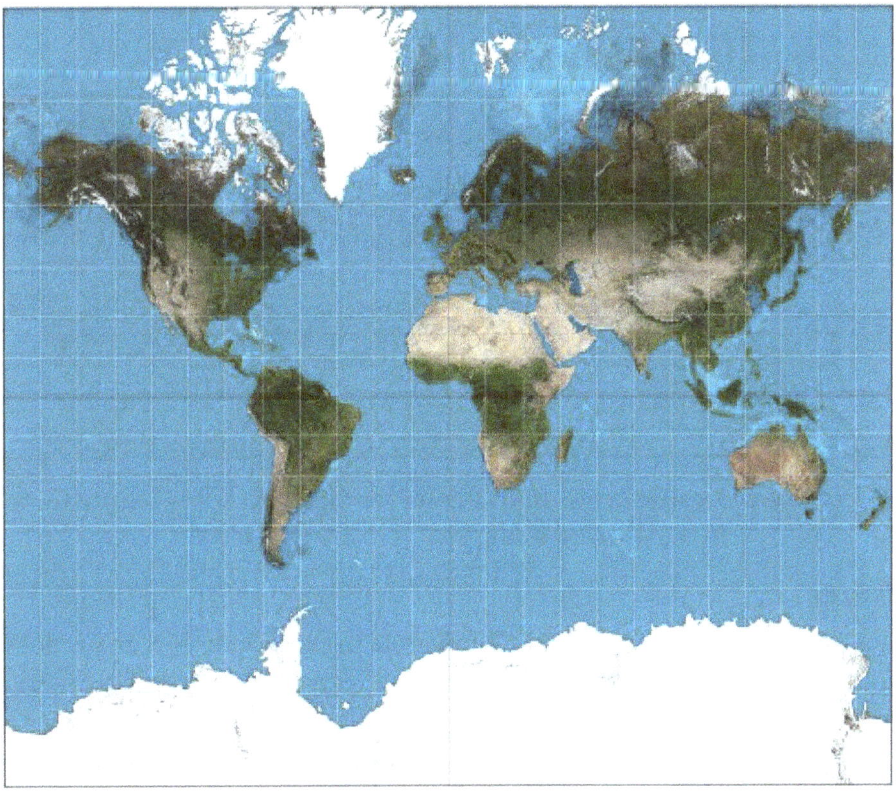

Figure 4.9 The Mercator Projection. Image is used under a CC BY-SA 3.0.

Conformal projections should be used if the main purpose of the map involves measuring angles or representing the shapes of features. They are very useful for navigation, topography (elevation), and weather maps.

A conformal projection will have distortion ellipses that vary substantially in size but are all the same circular shape. The consistent shapes indicate that conformal projections (like this Mercator projection of the world) preserve shapes and angles. This useful property accounts for the fact that conformal projections are almost always used as the basis for large scale surveying and mapping.

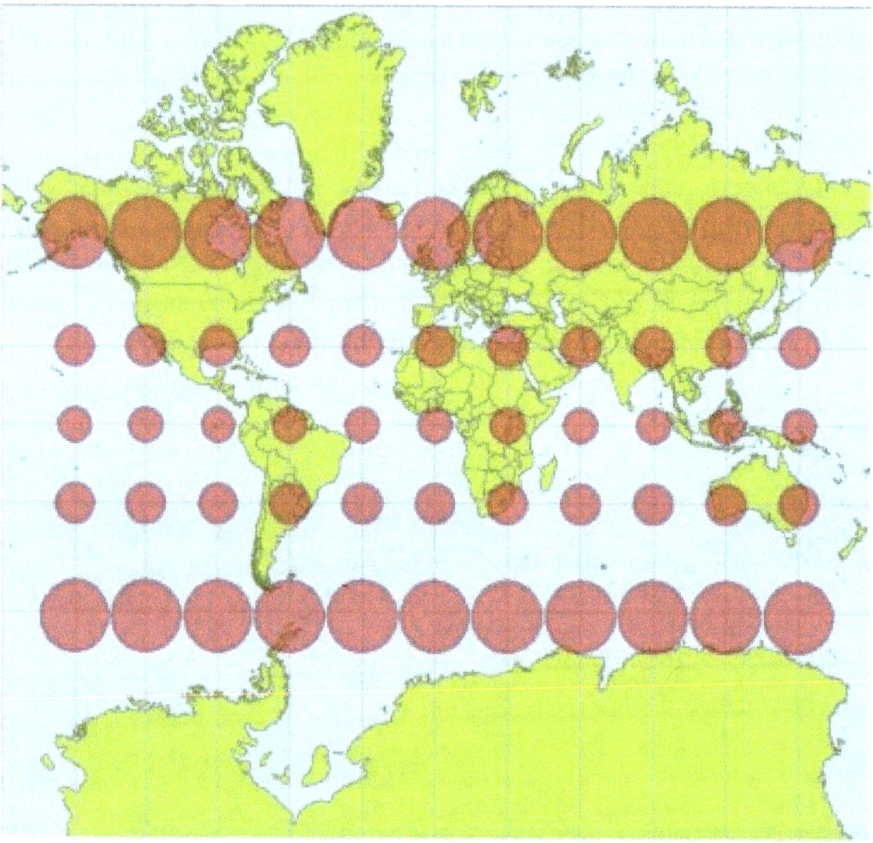

Figure 4.10 The Mercator Projection- Preserves Shape & Angle but Distorts Area. Image is used under a CC BY-SA 3.0. license.

Equal Area

For equal-area projections, the size of any area on the map is in true proportion to its size on the earth. In other words, countries' shapes may appear to be squished or stretched compared to what they look like on a globe, but their land area will be accurate relative to other landmasses. For example, in the Gall-Peters projection, the shape of Greenland is significantly altered, but the size of its area is correct in comparison to Africa. This type of projection is important for quantitative thematic data, especially in mapping density (an attribute over an area). For example, it would be useful in comparing the density of Syrian refugees in the Middle East or the amount of cropland in production.

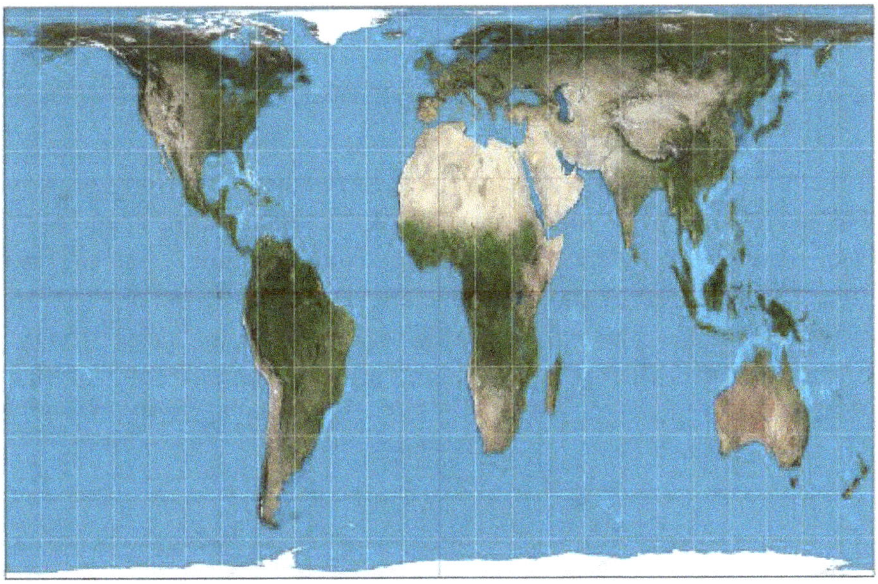

Figure 4.11 Galls-Peter Projection. Image is used under a CC BY-SA 3.0.

As we can see with an equal-area projection, however, the ellipses maintain the correct proportions in the sizes of areas on the globe but that their shapes are distorted. Equal-area projections are preferred for small-scale thematic mapping, especially when map users are expected to compare sizes of area features like countries and continents.

Figure 4.12 Gall-Peters Projection- Area is Preserved, But Shapes Are Heavily Distorted.Image is used under a CC BY-SA 3.0.

Equidistant

Equidistant projections, as the name suggests, preserve distance. This is a bit misleading because no projection can maintain relative distance between all places on the map.

Equidistant maps are able, however, to preserve distances along a few specified lines. For example, on the Azimuthal Equidistant projection, all points are the proportionally correct distance and direction from the center point. This type of projection would be useful visualizing airplane flight paths from one city to several other cities or in mapping an earthquake epicenter. Azimuthal projections preserve distance at the cost of distorting shape and area to some extent. The flag of the United Nations contains an example of a polar azimuthal equidistant projection.

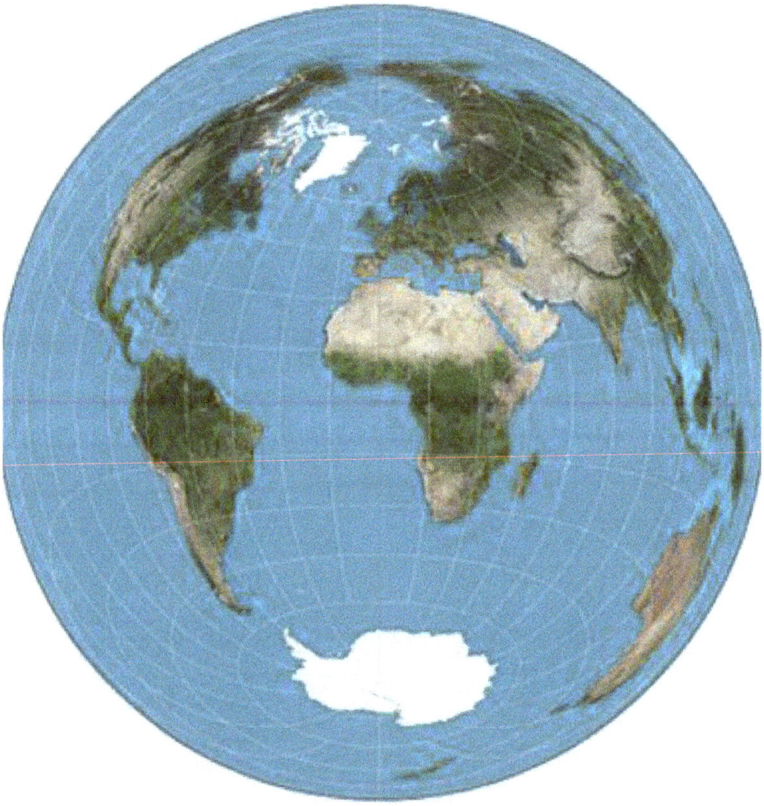

Figure 4.13 Azimuthal Equidistant Projection- All Points are the Proportionally Correct Distances from a Central Point. Image is used under a CC BY-SA 3.0.

Compromise, Interrupted, & Artistic Projections

Some projections, including the Robinson projection, strike a balance between the different map properties. In other words, instead of preserving the shape, area, or distance, they try to avoid extreme distortion of any of these properties. This type of projection would be useful for a general-purpose world map.

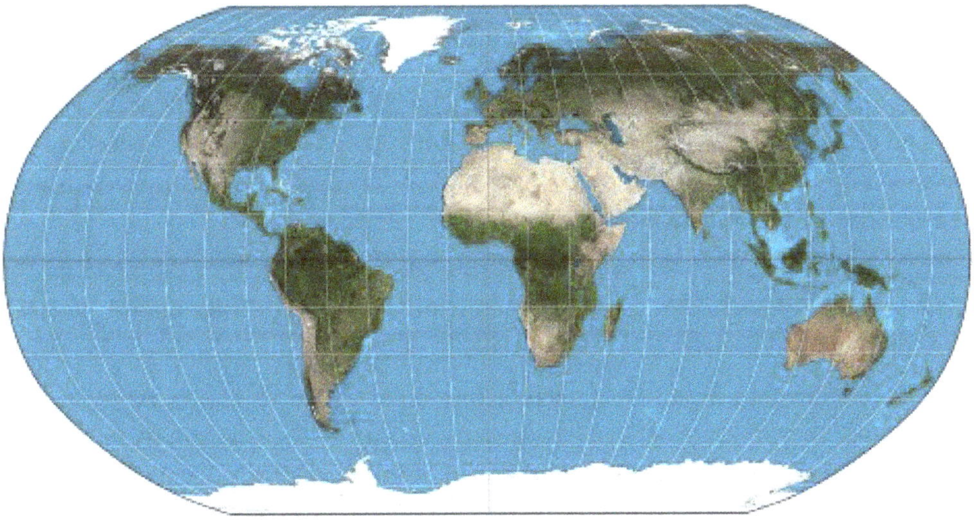

Figure 4.14 Robinson Projection of the World. Image is used under a CC BY-SA 3.0.

Compromise projections preserve not one property but instead seek a compromise that minimizes distortion of all kinds, as with the Robinson projection, which is often used for small- scale thematic maps of the world.

Other projections deal with the challenge of making the 3D globe flat by tearing the earth in strategic places. Interrupted projections such as the interrupted Goode Homolosine projection represent the earth in lobes, reducing the amount of shape and area distortion near the poles. The projection was developed in 1923 by John Paul Goode to provide an alternative to the Mercator projection for portraying global areal relationships. The Interrupted Goode Homolosine preserves area (so it is equal-area or equivalent) but does not preserve shape (it is not conformal).

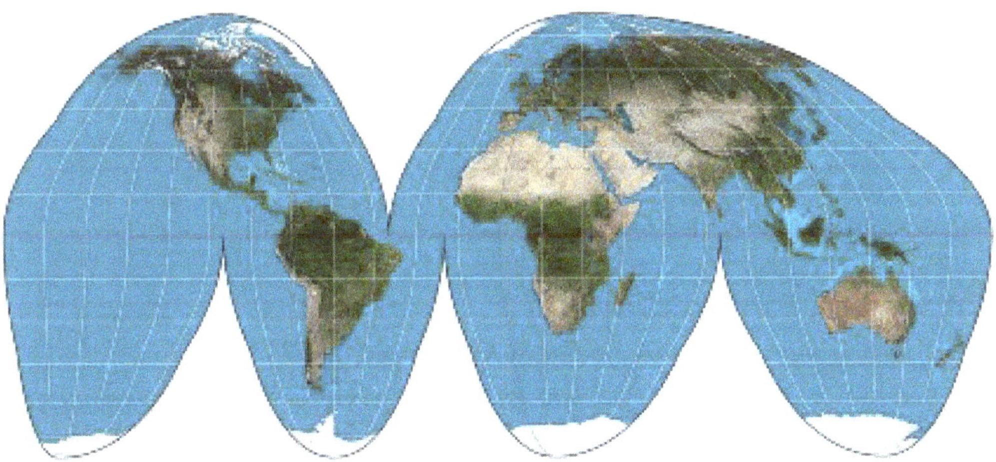

Figure 4.15 The Goodes Homolosine Projection of the World. Image is used under a CC BY-SA 3.0.

PUBLIC LAND SURVEY

The combination of a topographic map and this system can be used to locate features within a few acres and is a primary means of subdividing tracts of land for sale. The organization of the township-section system is based on the definition of baselines and principal meridians. The position of a baseline and meridian within a region may or may not coincide with latitude and longitude.

Townships are areas of 6 miles on a side (36 sq. mi), bordered on the east and west by range lines and the north and south by township lines. Each township is subdivided into 36 sections of 1 mile on each side.

When is it broken down even smaller into quadrants, which are where we come across ¼ mile increments, otherwise known as 40-acres. Ultimately this system was designed when man headed west, and each person was allotted a 40-acre parcel of land, and this was the best way to divide up the land since surveying at that time was too slow and expensive to complete.

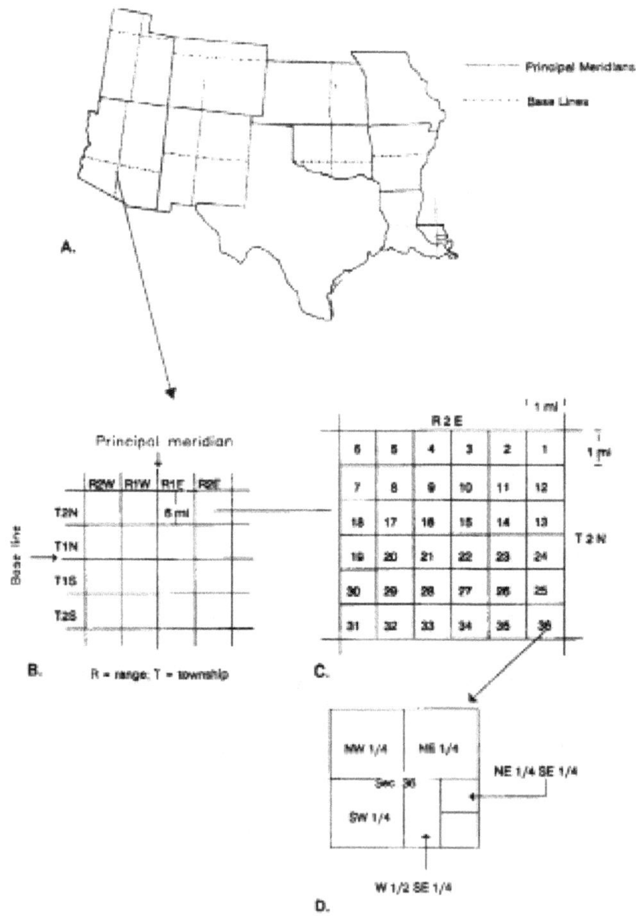

Figure 4.16 The Public Land Survey (USGS). Image is in the public domain.

SUMMARY

Students explored the concepts of scale, resolution, and projection. All maps also use a projection that can be formed from a developable surface and can preserve one or two properties at most.

In cartography, a map projection is a way to flatten a globe's surface into a plane to make a map. This requires a systematic transformation of the latitudes and longitudes of locations from the surface of the globe into locations on a plane. All projections of a sphere on a plane necessarily distort the surface in some way and to some extent. Depending on the purpose of the map, some distortions are acceptable, and others are not; therefore, different map projections exist to preserve some properties of the sphere-like body at the expense of other properties.

Latitude and longitude coordinates specify point locations within a coordinate system grid that is fitted to sphere or ellipsoid that approximates the Earth's shape and size. To display extensive geographic areas on a page or computer screen, as well as to calculate distances, areas, and other quantities most efficiently, it is necessary to flatten the Earth.

5 Chapter

EARTH-SUN RELATIONSHIPS: REASONS FOR THE SEASONS

Figure 5.1 Fall Colors at Lake Sabrina in Bishop, California. Image by Jeremy Patrich is under a CC-BY 4.0 license.

GOALS & OBJECTIVES OF THIS UNIT

- Describe how Earth's movements affect seasons and cause day and night.
- Identify the relationships between latitude, length of daylight, or night, as it pertains to incipient solar angles.
- Explain and identify the similarities and differences between solar and lunar eclipses.
- Describe the phases of the Moon and explain why they occur.
- Explain how movements of the Earth and Moon affect Earth's tides.

THE SUN AND THE EARTH SYSTEM

The solar system is made up of the Sun, the planets that orbit the Sun, their satellites, dwarf planets, and many, many small objects, like asteroids and comets. All of these objects move, and we can see these movements. We notice the Sun rises in the eastern sky in the morning and sets in the western sky in the evening. We observe different stars in the sky at different times of the year. When ancient people made these observations, they imagined that the sky was moving while the Earth stood still. In 1543, Nicolaus Copernicus proposed a radically different idea: The Earth and the other planets make regular revolutions around the Sun. He also suggested that the Earth rotates once a day on its axis. Copernicus' idea slowly gained acceptance and today we base our view of motions in the solar system on his work. We also now know that everything in the universe is moving.

Positions & Movements

The Earth rotates once on its axis about every 24 hours. If you were to look at Earth from the North Pole, it would be spinning counterclockwise. As the Earth rotates, observers on Earth see the Sun moving across the sky from east to west with the beginning of each new day. We often say that the Sun is "rising" or "setting", but it is the Earth's rotation that gives us the perception of the Sun rising or setting over the horizon. When we look at the Moon or the stars at night, they also seem to rise in the east and set in the west. Earth's rotation is also responsible for this. As Earth turns, the Moon and stars change position in our sky.

Another effect of Earth's rotation is that we have a cycle of daylight and darkness approximately every 24 hours. This is called a day. As Earth rotates, the side of Earth facing the Sun experiences daylight, and the opposite side (facing away from the Sun) experiences darkness or nighttime. Since the Earth completes one rotation in about 24 hours, this is the time it takes to complete one day-night cycle. As the Earth rotates, different places on Earth experience sunset and sunrise at a different time. As you move towards the poles, summer and winter days have different amounts of daylight hours in a day. For example, in the Northern Hemisphere, we begin summer on or around June 21st. At this point, the Earth's North Pole is pointed directly toward the Sun. Therefore, areas north of the equator experience longer days and shorter nights because the northern half of the Earth is pointed toward the Sun. Since the southern half of the Earth is pointed away from the Sun at that point, they have the opposite effect, longer nights and shorter days.

For people in the Northern Hemisphere, winter begins on or around December 21st. At this point, it is Earth's South Pole that is tilted toward the Sun, and so there are shorter days and longer nights for those who are north of the equator.

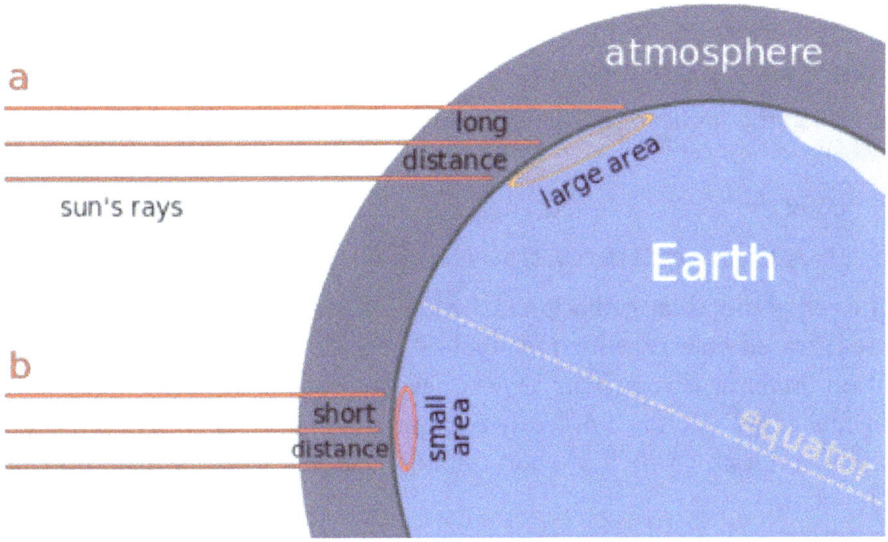

Figure 5.2 The Earth's Tilt on its Axis Leads to One Hemisphere Facing the Sun More Than the Other Hemisphere and Gives Rise to the Seasons. Image is under a Creative Commons Attribution-Share Alike 3.0 Unported license.

Energy from the Sun

The earth constantly tries to maintain an energy balance with the atmosphere. Most of the energy that reaches the Earth's surface comes from the Sun. About 44% of solar radiation is in the visible light wavelengths, but the Sun also emits infrared, ultraviolet, and other wavelengths. When viewed together, all of the wavelengths of visible light appear white. But a prism or water droplets can break the white light into different wavelengths so that separate colors appear.

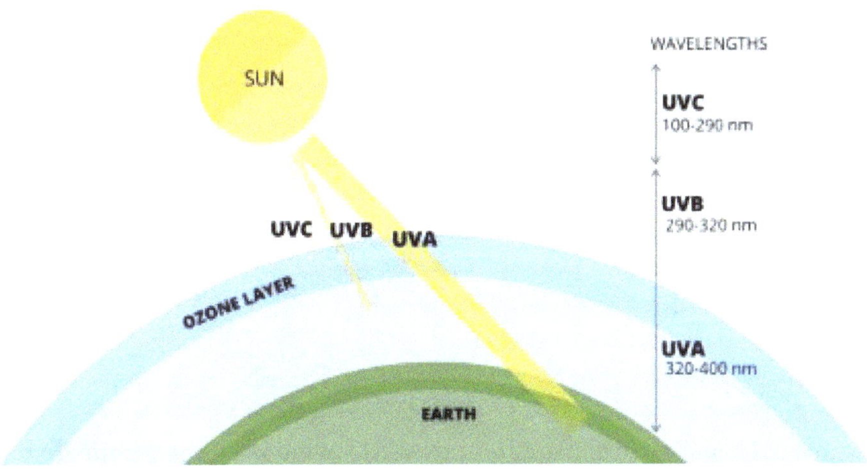

Figure 5.3 Diagram Showing the Three Types of Ultra Violet Light Emitted from the Sun. Image by Trudi Radtke team is used under a CC BY 4.0 license.

Of the solar energy that reaches the outer atmosphere, UV wavelengths have the greatest energy. Only about 7% of solar radiation is in the UV wavelengths. The three types are:

- UVC: the highest energy ultraviolet, does not reach the planet's surface at all.
- UVB: the second-highest energy, is also mostly stopped in the atmosphere.
- UVA: the lowest energy, travels through the atmosphere to the ground.

The remaining solar radiation is the longest wavelength, infrared. Most objects radiate infrared energy, which we feel as heat. Some of the wavelengths of solar radiation traveling through the atmosphere may be lost because they are absorbed by various gases. Ozone completely removes UVC, most UVB and some UVA from incoming sunlight. Oxygen, carbon dioxide, and water vapor also filter out some wavelengths.

THE GREENHOUSE EFFECT

The exception to Earth's temperature being in balance is caused by greenhouse gases. But first, the role of greenhouse gases in the atmosphere must be explained. Greenhouse gases warm the atmosphere by trapping heat.

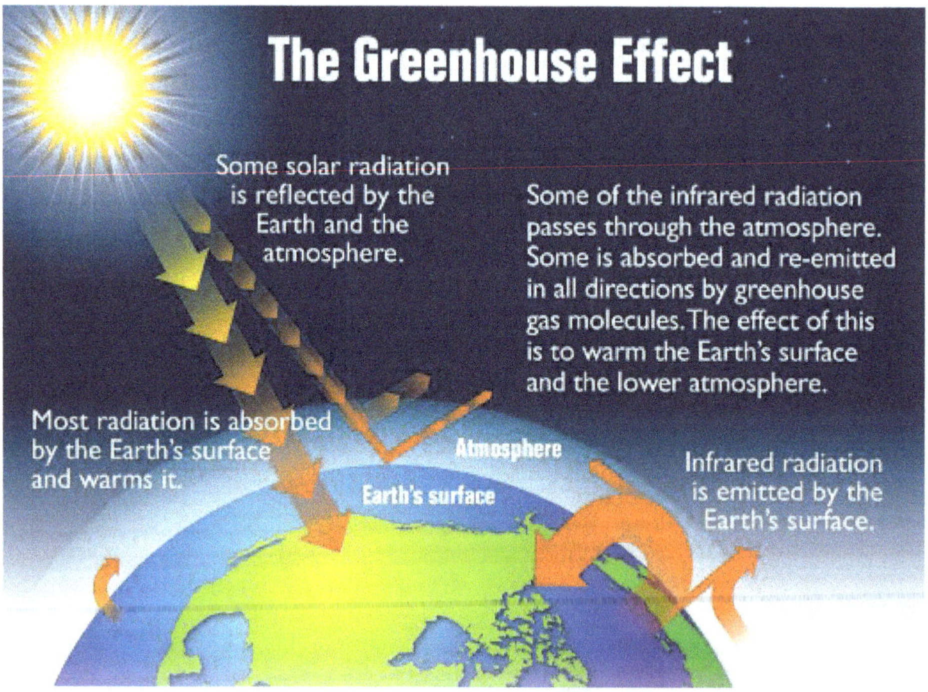

Figure 5.4 The Greenhouse Effect (NASA).Image is in the public domain.

Some of the heat radiation out from the ground is trapped by greenhouse gases in the troposphere. Like a blanket on a sleeping person, greenhouse gases act as insulation for the planet. The warming of the atmosphere because of insulation by greenhouse gases is called the greenhouse effect. Greenhouse gases are the component of the atmosphere that moderate Earth's temperatures.

Greenhouse gases include CO2, H2O, methane, O3, nitrous oxides (NO and NO2), and chlorofluorocarbons (CFCs). All are a normal part of the atmosphere except CFCs. The table below shows how each greenhouse gas naturally enters the atmosphere.

Different greenhouse gases have different abilities to trap heat. For example, one methane molecule traps 23 times as much heat as one CO2 molecule. One CFC-12 molecule (a type of CFC) traps 10,600 times as much heat as one CO2. Still, CO2 is a very important greenhouse gas because it is much more abundant in the atmosphere. Human activity has significantly raised the levels of many greenhouse gases in the atmosphere. Methane levels are about 2 ½ times higher as a result of human activity. Carbon dioxide has increased by more than 35%. CFCs have only recently existed.

What do you think happens as atmospheric greenhouse gas levels increase? More greenhouse gases trap more heat and warm the atmosphere. The increase or decrease of greenhouse gases in the atmosphere affect climate and weather the world over.

EARTH'S SEASONS

It is a common misconception that summer is warm, and winter is cold because the Sun is closer to Earth in the summer and farther away from it during the winter. Remember that seasons are caused by the 23.5° tilt of Earth's axis of rotation and Earth's yearly revolution around the Sun. This results in one part of the Earth being more directly exposed to rays from the Sun than the other part. The part tilted away from the Sun experiences a cool season, while the part tilted toward the Sun experiences a warm season. Seasons change as the Earth continues its revolution, causing the hemisphere tilted away from or towards the Sun to change accordingly. When it is winter in the Northern Hemisphere, it is summer in the Southern Hemisphere, and vice versa.

Northern Hemisphere Summer

The North Pole is tilted towards the Sun and the Sun's rays strike the Northern Hemisphere more directly in summer. At the summer solstice, which is around June 21st or 22nd, the Sun's rays hit the Earth most directly along the Tropic of Cancer (23.5° N); that is, the angle of incidence of the sun's rays there is zero (the angle of incidence is the deviation in the angle of an incoming ray from straight on). When it is the summer solstice in the Northern Hemisphere, it is the winter solstice in the Southern Hemisphere.

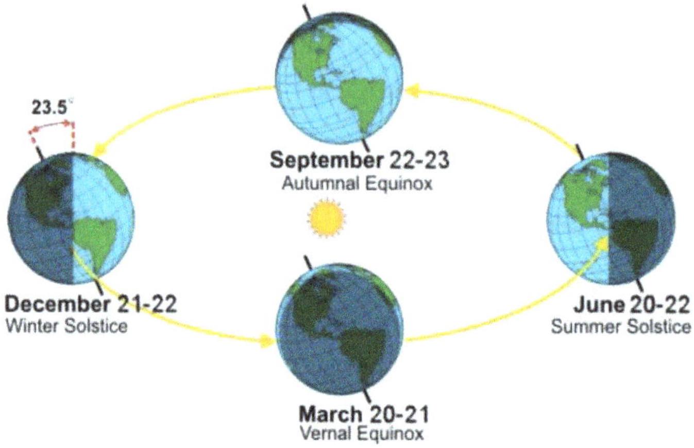

Figure 5.5 Seasons Diagram, Note the Tilt and Circle of Illumination for Each Season. Image is in the public domain.

Northern Hemisphere Winter

The Winter solstice for the Northern Hemisphere happens on or around December 21st or the 22nd. The tilt of Earth's axis points away from the Sun. Light from the Sun is spread out over a larger area, so that area isn't heated as much. With fewer daylight hours in winter, there is also less time for the Sun to warm the area. When it is winter in the Northern Hemisphere, it is summer in the Southern Hemisphere.

Equinox

Halfway between the two solstices, the Sun's rays shine most directly at the equator, called an "equinox." The daylight and nighttime hours are exactly equal on an equinox. The autumnal equinox happens on or around September 22nd or the 23rd and the vernal or spring equinox happens on or around March 21st or 22nd in the Northern Hemisphere.

Analemma

In astronomy, an analemma is a diagram showing the position of the Sun in the sky, as seen from a fixed location on Earth at the same mean solar time, as that position varies over the course of a year. The north-south component of the analemma results from the change in the Sun's declination due to the tilt of Earth's axis of rotation. The east-west component results from the nonuniform rate of change of the Sun's right ascension, governed by combined effects of Earth's axial tilt and orbital eccentricity.

An analemma can be traced by plotting the position of the Sun as viewed from a fixed position on Earth at the same clock time every day for an entire year, or by plotting a graph of the Sun's declination against the equation of time. The resulting curve resembles a long, slender figure- eight with one lobe much larger than the other. This curve is commonly printed on terrestrial globes, usually in the eastern Pacific Ocean, the only large tropical region with very little land. It is possible, though challenging, to photograph the analemma, by leaving the camera in a fixed position for an entire year and snapping images on 24-hour intervals.

The long axis of the figure, the line segment joining the northernmost point on the analemma to the southernmost, is bisected by the celestial equator, to which it is approximately perpendicular, and has a "length" of twice the obliquity of the ecliptic, e.g., about 47°. The component along this axis of the Sun's apparent motion is a result of the familiar seasonal variation of the declination of the Sun through the year. The "width" of the figure is due to the equation of time, and its angular extent is the difference between the greatest positive and negative deviations of local solar time from the local mean time when this time-difference is related to the angle at the rate of 15° per hour, e.g., 360° in 24 hours. The difference in the size of the lobes of the figure-eight form arises mainly from the fact that the perihelion and aphelion occur far from equinoxes. They also occur a mere couple of weeks after solstices, which in turn causes a slight tilt of the figure eight and its minor lateral asymmetry.

Figure 5.6 Diagram or Earth during the Aphelion (Away) & Perihelion (Near). Image by Trudi Radtke is used under a CC-BY 4.0 license.

Three parameters affect the size and shape of the analemma, which are eccentricity, obliquity, and the angle between the apse line and the line of solstices. Viewed from an object with a perfectly circular orbit and no axial tilt, the Sun would always appear at the same point in the sky at the same time of day throughout the year and the analemma would be a dot. For an object with a circular orbit but significant axial tilt, the analemma would be a figure eight with northern and southern lobes equal in size. For an object with an eccentric orbit but no axial tilt, the analemma would be a straight east-west line along the celestial equator.

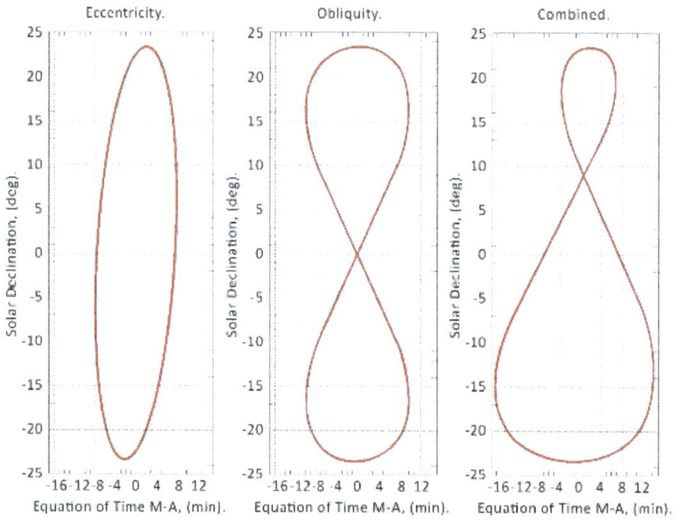

Figure 5.7 The Three Variation of Analemmas, Eccentricity, Obliquity & Combined. Image by Anthony Flores is used under a CC-BY-4.0 license.

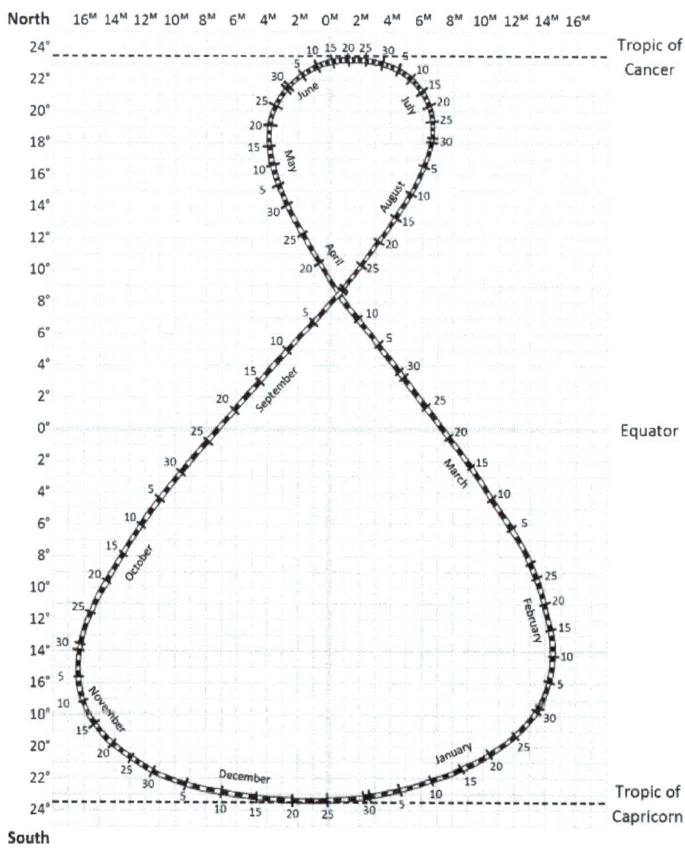

Figure 5.8 The Analemma. Image by Anthony Flores is used under a CC-BY-4.0 license.

The north-south component of the analemma shows the Sun's declination, its latitude on the celestial sphere, or the latitude on the Earth at which the Sun is directly overhead. The east- west component shows the equation of time or the difference between solar time and local meantime. This can be interpreted as how "fast" or "slow" the Sun (or a sundial) is compared to clock time. It also shows how far west or east the Sun is, compared with its mean position. The analemma can be considered as a graph in which the Sun's declination and the equation of time are plotted against each other. In many diagrams of the analemma, a third dimension, that of time, is also included, shown by marks that represent the position of the Sun at various, fairly closely spaced, dates throughout the year.

SUMMARY

As the Earth rotates on its axis and revolves around the Sun, several different effects are produced.

The summer solstice is the longest day of the year, and the winter solstice is the shortest. The equinox is the time when the day and night are the same numbers of hours.

The reasons for the seasons are beyond the length of daylight, it also includes the intensity and duration of daylight, as well as Earth's physical distance from the Sun.

The analemma is caused by the tilt of the Earth's axis as it rotates and the elliptical shape of Earth's orbit around the Sun.

There are measurable distances between the Earth and Sun throughout the year, identified as the perihelion and aphelion.

6 Chapter

EARTH'S ATMOSPHERE

Figure 6.1 At 12,200ft in the Ancient Bristlecone Pine Forest. The White Mountains, California.

Goals & Objectives of this unit

- Understand the significance of the atmosphere.
- Describe the composition of the atmospheric gasses.
- Explain the major layers of the atmosphere and their importance.
- Analyze the relationships between energy, temperature, and heat.
- Identify and explore the Earth's solar budget and identify the values of albedo.
- Describe how heat is transferred around the planet.

THE SIGNIFICANCE OF THE ATMOSPHERE

Earth's atmosphere is a thin blanket of gases and tiny particles called air. We are most aware of air when it moves and creates wind. All living things need some of the gases in the air for life support. Without an atmosphere, Earth would likely be just another lifeless rock. Earth's atmosphere, along with the abundant liquid water at Earth's surface, is the keys to our planet's unique place in the solar system. Much of what makes Earth exceptional depends on the atmosphere. Let's consider some of the reasons we are lucky to have an atmosphere.

Indispensable for Life on Earth

Without the atmosphere, Earth would look a lot more like the Moon. Atmospheric gases, especially carbon dioxide (CO_2) and oxygen (O_2) are extremely important for living organisms. How does the atmosphere make life possible? How does life alter the atmosphere?

In photosynthesis, plants use CO2 and create O2. Photosynthesis is responsible for nearly all of the oxygen currently found in the atmosphere. By creating oxygen and food, plants have made an environment that is favorable for animals. In respiration, animals use oxygen to convert sugar into food energy they can use. Plants also go through respiration and consume some of the sugars they produce.

Along with the oceans, the atmosphere keeps Earth's temperatures within an acceptable range. Greenhouse gases trap heat in the atmosphere so they help to moderate global temperatures. Without an atmosphere with greenhouse gases, Earth's temperatures would be frigid at night and scorching during the day. Important greenhouse gases include carbon dioxide, methane, water vapor, and ozone.

ATMOSPHERIC GASSES

Nitrogen and oxygen together make up 99% of the planet's atmosphere. The rest of the gases are minor components but are sometimes very important. Humidity is the amount of water vapor in the air. Humidity varies from place to place and season to season. This fact is obvious if you compare a summer day in Atlanta, Georgia, where humidity is high, with a winter day in Phoenix, Arizona, where humidity is low. When the air is very humid, it feels heavy or sticky.

Dry air usually feels more comfortable. Higher humidity is found around the equatorial regions because air temperatures are higher and warm air can hold more moisture than cooler air. Of course, humidity is lower near the Polar Regions because the air temperature is lower.

Some of what are in the atmosphere is not gas. Particles of dust, soil, fecal matter, metals, salt, smoke, ash, and other solids make up a small percentage of the atmosphere. Particles provide starting points (or nuclei) for water vapor to condense on and form raindrops.

Table 6.1 Showing Atmospheric Gasses as a Percent by Volume

Gas	Symbol	Volume (%)
Nitrogen	N2	78.0840
Oxygen	O2	20.9480
Argon	A	0.9340
Carbon Dioxide	CO2	0.0314
Neon	Ne	0.0018
Helium	He	0.0005
Hydrogen	H2	<0.0001

Atmospheric Pressure & Density

The atmosphere has different properties at different elevations above sea level or altitudes. The air density (the number of molecules in a given volume) decreases with increasing altitude. This is why people who climb tall mountains, such as Mt. Everest, have to set up camp at different elevations to let their bodies get used to the decreased air. Why does air density decrease with altitude? Gravity pulls the gas molecules towards Earth's center. The pull of gravity is stronger closer to the center at sea level. Air is denser at sea level where the gravitational pull is greater. Gases at sea level are also compressed by the weight of the atmosphere above them. The force of the air weighing down over a unit of area is known as its atmospheric pressure.

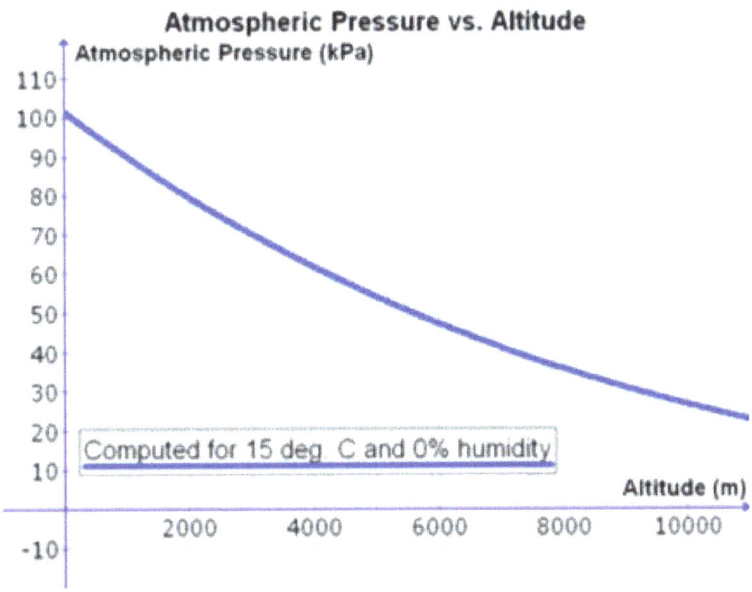

Figure 6.2 Diagram Explaining how Atmospheric Pressure Changes with Altitude. Image has been designated to the public domain under a CC0 1.0 Universal Public Domain Dedication.

The reason why we are not crushed by this weight is that the molecules inside our bodies are pushing outward to compensate. Atmospheric pressure is felt from all directions, not just from above.

At higher altitudes, the atmospheric pressure is lower, and the air is less dense than at higher altitudes. If your ears have ever "popped", you have experienced a change in air pressure. Gas molecules are found inside and outside your ears. When you change altitude quickly, like when an airplane is descending, your inner ear keeps the density of molecules at the original altitude. Eventually, the air molecules inside your ear suddenly move through a small tube in your ear to equalize the pressure. This sudden rush of air is felt as a popping sensation.

Although the density of the atmosphere changes with altitude, the composition stays the same with altitude, with one exception. In the ozone layer, at about 20 km to 40 km above the surface, there is a greater concentration of ozone molecules than in other portions of the atmosphere.

LAYERS OF THE ATMOSPHERE

The atmosphere is layered, corresponding with how the atmosphere's temperature changes with altitude. By understanding the way temperature changes with altitude, we can learn a lot about how the atmosphere works. While weather takes place in the lower atmosphere, interesting things, such as the beautiful aurora, happen higher in the atmosphere. Why does warm air rise? Gas molecules can move freely and if they are uncontained, as they are in the atmosphere, they can take up more or less space.

- When gas molecules are cool, they are sluggish and do not take up as much space. Withthe same number of molecules in less space, both air density and air pressure are higher.
- When gas molecules are warm, they move vigorously and take up more space. Airdensity and air pressure are lower.

Warmer, lighter air is more buoyant than the cooler air above it, so it rises. The cooler air then sinks, because it is denser than the air beneath it.

The property that changes most strikingly with altitude is air temperature. Unlike the change in pressure and density, which decreases with altitude, changes in air temperature are not regular. A change in temperature with distance is called a temperature gradient.

The atmosphere is divided into layers based on how the temperature in that layer changes with altitude, the layer's temperature gradient. The temperature gradient of each layer is different. In some layers, temperature increases with altitude, and others, it decreases. The temperature gradient in each layer is determined by the heat source of the layer. Most of the important processes of the atmosphere take place in the lowest two layers: the troposphere and the stratosphere.

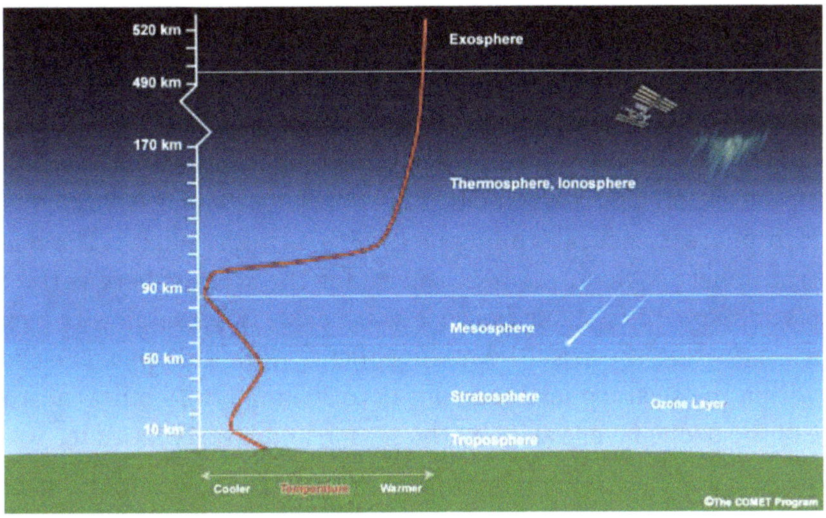

Figure 6.3 Layers of Earth's Atmosphere & Temperature Gradient. Image is in the public domain.

Troposphere

The temperature of the troposphere is highest near the surface of the Earth and decreases with altitude. On average, the temperature gradient of the troposphere is 6.5°C per 1,000 m (3.6°F per 1,000 ft.) of altitude. What is the source of heat for the troposphere? Earth's surface is a major source of heat for the troposphere, although nearly all of that heat comes from the Sun. Rock, soil, and water on Earth absorb the Sun's light and radiate it back into the atmosphere as heat. The temperature is also higher near the surface because of the greater density of gases.

The higher gravity causes the temperature to rise. Notice that in the troposphere warmer air is beneath cooler air. What do you think the consequence of this is? This condition is unstable.

The warm air near the surface rises and cool air higher in the troposphere sinks. The air in the troposphere does a lot of mixing. This mixing causes the temperature gradient to vary with time and place. The rising and sinking of air in the troposphere mean that all of the planet's weather takes place in the troposphere.

Sometimes there is a temperature inversion, air temperature in the troposphere increases with altitude and warm air sits over cold air. Inversions are very stable and may last for several days or even weeks. They form:

- Overland at night or in winter when the ground is cold. The cold ground cools the airthat sits above it, making this low layer of air denser than the air above it.

- Near the coast where cold seawater cools the air above it. When that denser air moves inland, it slides beneath the warmer air over the land.

Stratosphere

Ash and gas from a large volcanic eruption may burst into the stratosphere, the layer above the troposphere. Once in the stratosphere, it remains suspended there for many years because there is so little mixing between the two layers. Pilots like to fly in the lower portions of the stratosphere because there is little air turbulence. In the stratosphere, temperature increases with altitude. What is the heat source for the stratosphere? The direct heat source for the stratosphere is the Sun. The air in the stratosphere is stable because warmer, less dense air sits over cooler, denser air. As a result, there is little mixing of air within the layer. The ozone layer is found within the stratosphere between 15 to 30 km (9 to 19 miles) altitudes. The thickness of the ozone layer varies by the season and also by latitude.

The ozone layer is extremely important because ozone gas in the stratosphere absorbs most of the Sun's harmful ultraviolet (UV) radiation. Because of this, the ozone layer protects life on Earth. The high-energy UV light penetrates cells and damages DNA, leading to cell death (which we know as a bad sunburn). Organisms on Earth are not adapted to heavy UV exposure, which kills or damages them. Without the ozone layer to reflect UVC and UVB radiation, the most complex life on Earth would not survive long.

Mesosphere

Temperatures in the mesosphere decrease with altitude. Because there are few gas molecules in the mesosphere to absorb the Sun's radiation, the heat source is the stratosphere below. The mesosphere is extremely cold, especially at its top, about -90° C (-130° F).

The air in the mesosphere has extremely low density: 99.9% of the mass of the atmosphere is below the mesosphere. As a result, air pressure is very low. A person traveling through the mesosphere would experience severe burns from ultraviolet light since the ozone layer which provides UV protection is in the stratosphere below. There would be almost no oxygen for breathing. Stranger yet, an unprotected traveler's blood would boil at normal body temperature because the pressure is so low.

Thermosphere

The density of molecules is so low in the thermosphere that one gas molecule can go about 1 km before it collides with another molecule. Since so little energy is transferred, the air feels very cold. Within the thermosphere is the ionosphere. The ionosphere gets its name from the solar radiation that ionizes gas molecules to create a positively charged ion and one or more negatively charged electrons. The freed electrons travel within the ionosphere as electric currents. Because of the free ions, the ionosphere has many interesting characteristics. At night, radio waves bounce off the ionosphere and back to Earth. This is why you can often pick up an AM radio station far from its source at night. The Van Allen radiation belts are two doughnut-shaped zones of highly charged particles that are located beyond the atmosphere in the magnetosphere. The particles originate in solar flares and fly to Earth on the solar wind.

Once trapped by Earth's magnetic field, they follow along the field's magnetic lines of force. These lines extend from above the equator to the North Pole and also to the South Pole then return to the equator.

When massive solar storms cause the Van Allen belts to become overloaded with particles, the result is the most spectacular feature of the ionosphere or the nighttime aurora. The particles spiral along magnetic field lines toward the poles. The charged particles energize oxygen and nitrogen gas molecules, causing them to light up. Each gas emits a particular color of light.

ATMOSPHERIC HEAT, ENERGY & MOTION

Energy travels through space or material. This is obvious when you stand near a fire and feel its warmth or when you pick up the handle of a metal pot even though the handle is not sitting directly on the hot stove. Invisible energy waves can travel through air, glass, and even the vacuum of outer space. These waves have electrical and magnetic properties, so they are called electromagnetic waves. The transfer of energy from one object to another through electromagnetic waves is known as radiation. Different wavelengths of energy create different types of electromagnetic waves.

- The wavelengths humans can see are known as "visible light." These wavelengths
- appear to us as the colors of the rainbow. What objects can you think of that radiate visible light? Two include the Sun and a light bulb.
- The longest wavelengths of visible light appear red. Infrared wavelengths are longer
- than visible red. Snakes can see infrared energy. We feel infrared energy as heat.
- Wavelengths that are shorter than violet are called ultraviolet.

Can you think of some objects that appear to radiate visible light, but do not? The moon and the planets do not emit light of their own; they reflect the light of the Sun. Reflection is when light (or another wave) bounces back from a surface.

Albedo is a measure of how well a surface reflects light. A surface with high albedo reflects a large percentage of light. A snowfield has a high albedo.

One important fact to remember is that energy cannot be created or destroyed, it can only be changed from one form to another. This is such a fundamental fact of nature that it is a law: the law of conservation of energy.

In photosynthesis, for example, plants convert solar energy into chemical energy that they can use. They do not create new energy. When energy is transformed, some nearly always becomes heat. Heat transfers between materials easily, from warmer objects to cooler ones. If no more heat is added, eventually all of the material will reach the same temperature.

Temperature

Temperature is a measure of how fast the atoms in a material are vibrating. High-temperature particles vibrate faster than lo- temperature particles. Rapidly vibrating atoms smash together, which generates heat. As a material cools down, the atoms vibrate more slowly and collide less frequently. As a result, they emit less heat. What is the difference between heat and temperature?

- Temperature measures how fast a material's atoms are vibrating.
- Heat measures the material's total energy.
- Which has higher heat, and which has a higher temperature: a candle flame or a bathtub full of hot water?
- The flame has a higher temperature, but less heat, because the hot region is very small.
- The bathtub has a lower temperature but contains much more heat because it has many more vibrating atoms. The bathtub has greater total energy.

Heat

Heat is taken in or released when an object changes state, or changes from a gas to a liquid, or a liquid to a solid. This heat is called latent heat. When a substance changes state, latent heat is released or absorbed. A substance that is changing its state of matter does not change temperature. All of the energy that is released or absorbed goes toward changing the material's state.

For example, imagine a pot of boiling water on a stove burner: that water is at 100° C (212° F). If you increase the temperature of the burner, more heat enters the water. The water remains at its boiling temperature, but the additional energy goes into changing the water from liquid to gas. With more heat, the water evaporates more rapidly. When water changes from a liquid to a gas it takes in heat. Since evaporation takes in heat, this is called evaporative cooling.

Evaporative cooling is an inexpensive way to cool homes in hot, dry areas. Substances also differ in their specific heat, the amount of energy needed to raise the temperature of one gram of the material by 1.0° C (1.8° F). Water has a very high specific heat, which means it takes a lot of energy to change the temperature of the water. Let's compare a puddle and asphalt, for example. If you are walking barefoot on a sunny day, which would you rather walk across, the shallow puddle, or an asphalt parking lot? Because of its high specific heat, the water stays cooler than the asphalt, even though it receives the same amount of solar radiation.

Atmospheric Pressures & Winds

A few basic principles go a long way toward explaining how and why air moves: Warm air rising creates a low-pressure zone at the ground. Air from the surrounding area is sucked into the space left by the rising air. Air flows horizontally at top of the troposphere; horizontal flow is called advection. The air cools until it descends. When the air reaches the ground, it creates a high-pressure zone. Air flowing from areas of high pressure to low pressure creates winds.

Warm air can hold more moisture than cold air. Air moving at the bases of the three major convection cells in each hemisphere north and south of the equator creates the global wind belts.

Within the troposphere are convection cells. Air that moves horizontally between high- and low-pressure zones makes wind. The greater the pressure difference between the pressure zones the faster the wind flow. Convection in the atmosphere creates the planet's weather.

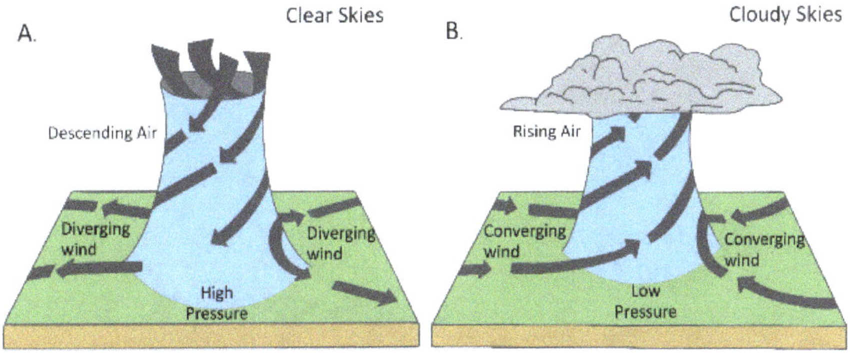

Figure 6.4 Diagram Explaining Both Low & High Pressure in the Northern Hemisphere. Image by COC OER team is used under a CC-BY-4.0 license.

When warm air rises and cools in a low-pressure zone, it may not be able to hold all the water it contains as vapor. Some water vapor may condense to form clouds or precipitation. When cool air descends, it warms. Since it can then hold more moisture, the descending air will evaporate water on the ground. Air moving between large high- and low-pressure systems create the global wind belts that profoundly affect regional climate. Smaller pressure systems create localized winds that affect the weather and climate of a local area.

Local Winds

Local winds result from air moving between small low and high-pressure systems. High- and low-pressure cells are created by a variety of conditions. Some local winds have very important effects on the weather and climate of some regions.

Land & Sea Breezes

Since water has a very high specific heat, it maintains its temperature well. As water is heated and cools more slowly than land. If there is a large temperature difference between the surface of the sea (or a large lake) and the land next to it, high- and low-pressure regions form. This creates local winds. Sea breezes blow from the cooler ocean over the warmer land in summer. Where is the high-pressure zone and where is the low-pressure zone? Sea breezes blow at about 10 to 20 km (6 to 12 miles) per hour and lower air temperature much as 5° to 10° C (9 to 18°F).

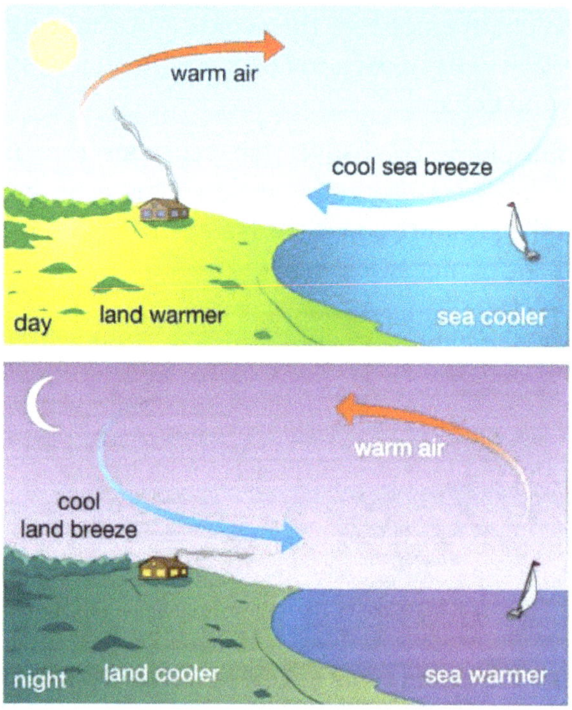

Figure 6.5 Land & Sea Breezes. Image by Encyclopedia Britannica used appropriately under Encyclopedia Britannica non-commercial terms of use.

Monsoonal Winds

Monsoon winds are larger-scale versions of land and sea breezes; they blow from the sea onto the land in summer and from the land onto the sea in winter. Monsoon winds occur where very hot summer lands are next to the sea. Thunderstorms are common during monsoons. The most important monsoon in the world occurs each year over the Indian subcontinent. More than two billion residents of India and southeastern Asia depend on monsoon rains for their drinking and irrigation water. Back in the days of sailing ships, seasonal shifts in the monsoon winds carried goods back and forth between India and Africa.

Mountain & Valley Breezes

Temperature differences between mountains and valleys create mountain and valley breezes. During the day, air on mountain slopes is heated more than air at the same elevation over an adjacent valley. As the day progresses, warm air rises and draws the cool air up from the valley, creating a valley breeze. At night the mountain slopes cool more quickly than the nearby valley, which causes a mountain breeze to flow downhill.

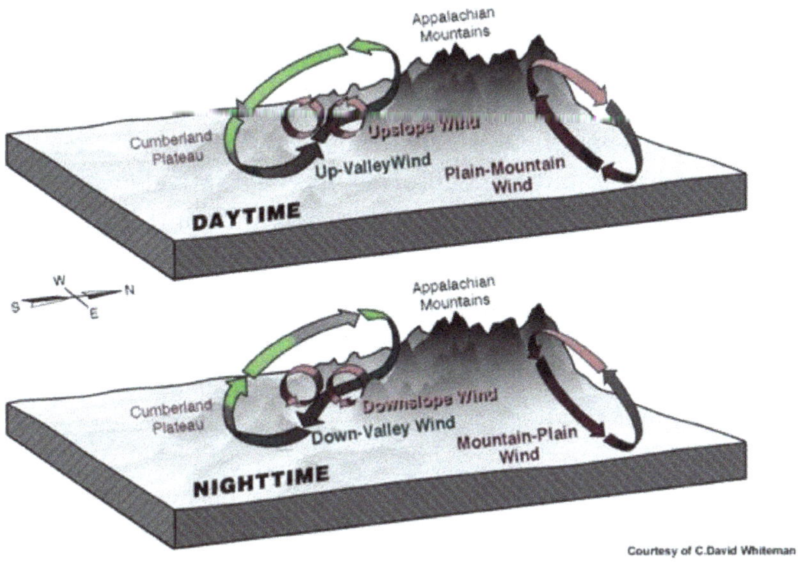

Figure 6.6 Mountain & Valley Breezes. Image is in the public domain.

Katabatic Winds

Katabatic winds move up and down slopes, but they are stronger mountain and valley breezes. Katabatic winds form over a high land area, like a high plateau. The plateau is usually surrounded on almost all sides by mountains. In winter, the plateau grows cold. The air above the plateau grows cold and sinks from the plateau through gaps in the mountains. Wind speeds depend on the difference in air pressure over the plateau and the surroundings. Katabatic winds form over many continental areas. Extremely cold katabatic winds blow over Antarctica and Greenland.

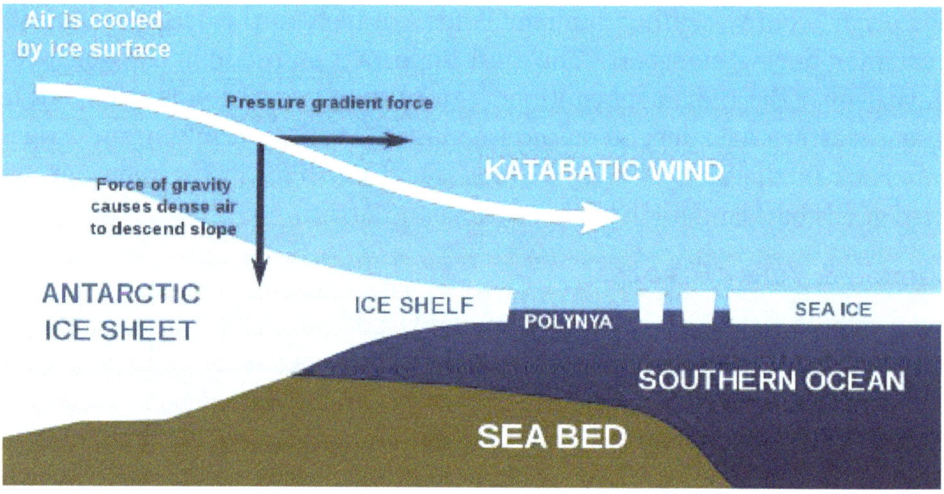

Figure 6.7 Katabatic Winds. Image is used under a Attribution-Share Alike 4.0 International license.

Chinook Winds

Chinook winds develop when air is forced up over a mountain range. This takes place, for example, when the westerly winds bring air from the Pacific Ocean over the Sierra Nevada Mountains in California. As the relatively warm, moist air rises over the windward side of the mountains, it cools and contracts. If the air is humid, it may form clouds and drop rain or snow.

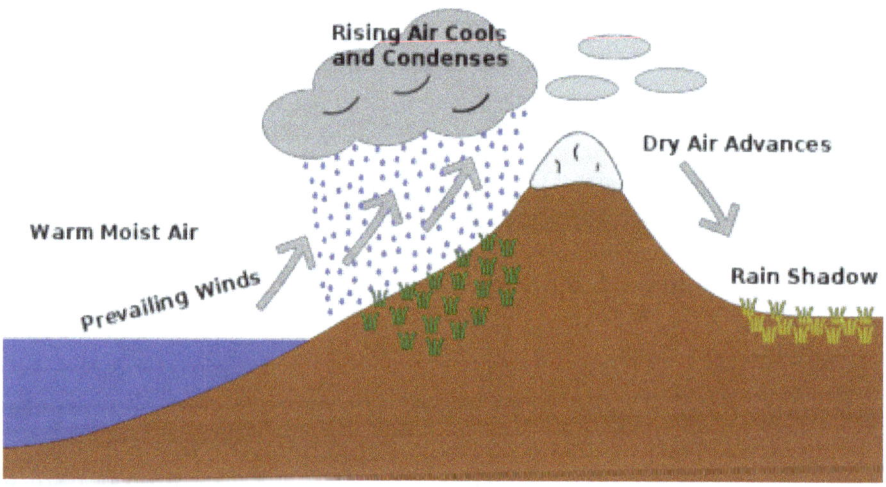

Figure 6.8 Rainshadow Effect. Image by Anders Einar Hilden has been released in the public domain.

When the air sinks on the leeward side of the mountains, it forms a high-pressure zone. The windward side of a mountain range is the side that receives the wind; the leeward side is the side where air sinks. The descending air warms and creates strong, dry winds. Chinook winds can raise temperatures more than 20°C (36°F) in an hour and they rapidly decrease humidity. If precipitation falls as the air rises over the mountains, the air will be dry as it sinks on the leeward side. This dry, sinking air causes a rain shadow effect, which creates many of the world's deserts.

Santa Ana Winds

Santa Ana winds are created in the late fall and winter when the Great Basin east of the Sierra Nevada cools, creating a high-pressure zone. The high-pressure forces the winds downhill and in a clockwise direction in the Northern Hemisphere (because of the Coriolis effect). The air pressure rises, so temperature rises and humidity falls. The winds blow across the Southwestern deserts and then race downhill and westward toward the ocean. Air is forced through canyons cutting the San Gabriel and San Bernardino mountains. The Santa Ana winds often arrive at the end of California's long summer drought season. The hot, dry winds dry out the landscape even more. If a fire starts, it can spread quickly, causing large-scale devastation.

Figure 6.9 The Santa Ana Winds-- Note the Direction of Flow Being Eastward. Image by NASA is in the public domain.

Desert Winds

High summer temperatures on the desert create high winds, which are often associated with monsoon storms. Desert winds pick up dust because there is not as much vegetation to hold down the dirt and sand. A haboob forms in the downdrafts on the front of a thunderstorm, as the ground becomes so hot that the air above it heats and rises. Air flows into the low pressure and begins to spin. Dust devils are small and short-lived, but they may cause damage.

Figure 6.10 A Haboob in Texas in 1935. Image is in the public domain.

GLOBAL ATMOSPHERIC CIRCULATIONS

Because more solar energy hits the equator, the air warms and forms a low-pressure zone. At the top of the troposphere, half moves toward the North Pole and half toward the South Pole. As it moves along the top of the troposphere it cools. The cool air is dense and when it reaches a high-pressure zone it sinks to the ground. The air is sucked back toward the low pressure at the equator. This describes the convection cells north and south of the equator. If the Earth did not rotate, there would be one convection cell in the Northern Hemisphere and one in the south with the rising air at the equator and the sinking air at each pole. But because the planet does rotate, the situation is more complicated. The planet's rotation means that the Coriolis effect must be taken into account. Let's look at atmospheric circulation in the Northern Hemisphere as a result of the Coriolis effect. Air rises at the equator, but as it moves toward the pole at the top of the troposphere, it deflects to the right. Remember that it just appears to deflect to the right because the ground beneath it moves. At about 30°N latitude, the air from the equator meets air flowing toward the equator from the higher latitudes. This air is cool because it has come from higher latitudes. Both batches of air descend, creating a high- pressure zone. Once on the ground, the air returns to the equator. This convection cell is called the Hadley Cell and is found between 0° and 30° N.

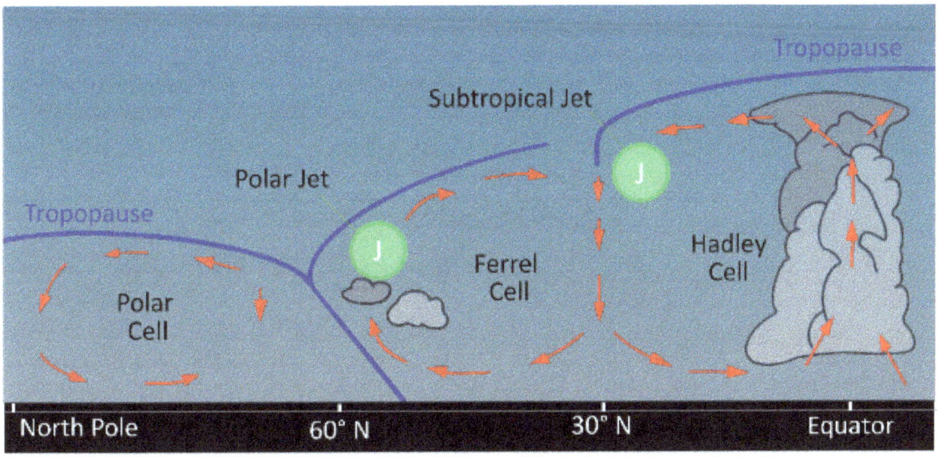

Figure 6.11 The Three Cell Model or Atmospheric Circulation. Image by Anthony Flores is used under a CC-BY-4.0 license.

There are two more convection cells in the Northern Hemisphere. The Ferrel cell is between 30°N and 50° to 60°N. This cell shares its southern, descending side with the Hadley cell to its south. Its northern rising limb is shared with the Polar cell located between 50° N to 60° N and the North Pole, where cold air descends.

There are three mirror image circulation cells in the Southern Hemisphere. In that hemisphere, the Coriolis Effect makes objects appear to deflect to the left. Ultimately, because there are three large-scale convection cells in the Northern Hemisphere and are repeated in the Southern Hemisphere, the model to understand these patterns is called the three-cell model.

Global Wind Patterns

Global winds blow in belts encircling the planet. The global wind belts are enormous, and the winds are relatively steady. These winds are the result of air movement at the bottom of the major atmospheric circulation cells, where the air moves horizontally from high to low pressure. Technology today allows anyone to see global wind patterns in real-time, such as the Earth Wind Map. Take a look at the Earth Wind Map and determine what patterns you can see occurring in the atmosphere in real-time. Are low-pressure systems rotating counterclockwise in the Northern Hemisphere? Are high-pressure systems rotating clockwise in the Northern Hemisphere? Can you see the global wind patterns over the Atlantic and Pacific Oceans? Also, notice how the winds flow faster over water than over continents because of land friction.

In the Hadley cell, air should move north to south, but it is deflected to the right by the Coriolis effect. So, the air blows from northeast to southwest. This belt is the trade winds, so-called because at the time of sailing ships they were good for trade.

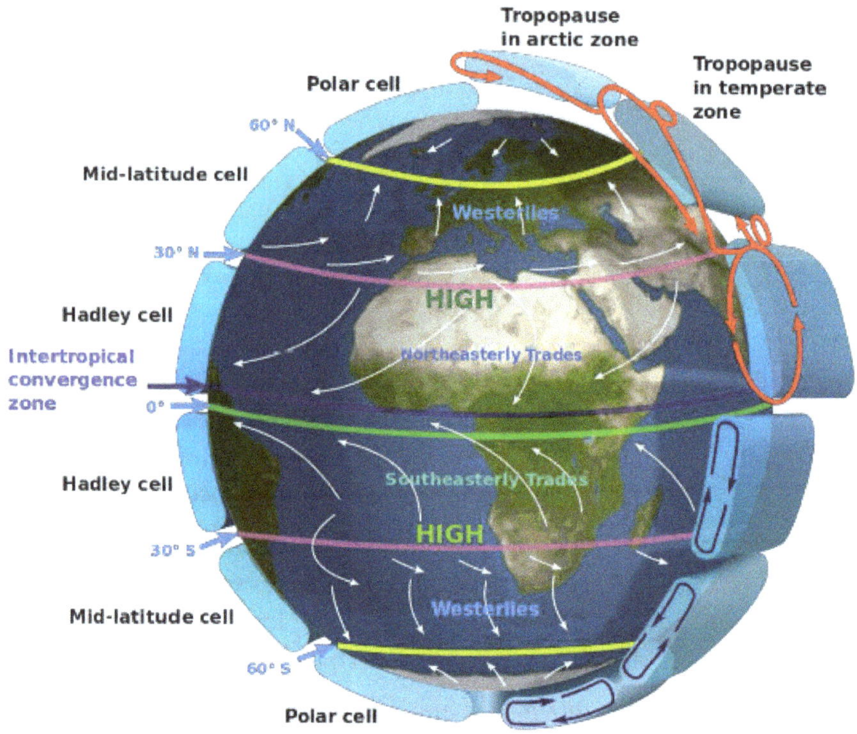

Figure 6.12 Global Winds & Atmospheric Circulation Model. Image is used under a Attribution-Share Alike 3.0 Unported license.

In the Ferrel cell, air should move south to north, but the winds blow from the southwest. This belt is the westerly winds or westerlies. Why do you think a flight across the United States from San Francisco to New York City takes less time than the reverse trip?

Finally, in the Polar cell, the winds travel from the northeast and are called the polar easterlies. The wind belts are named for the directions from which the winds come. The westerly winds, for example, blow from west to east. These names hold for the winds in the wind belts of the Southern Hemisphere as well.

The Polar Fronts & Jet Streams

The polar front is the junction between the Ferrel and Polar cells. At this low-pressure zone, relatively warm, moist air of the Ferrel Cell runs into relatively cold, dry air of the Polar cell. The weather where these two meet is extremely variable, typical of much of North America and Europe.

The polar jet stream is found high up in the atmosphere where the two cells come together. A jet stream is a fast-flowing river of air at the boundary between the troposphere and the stratosphere. Jet streams form where there is a large temperature difference between two air masses.

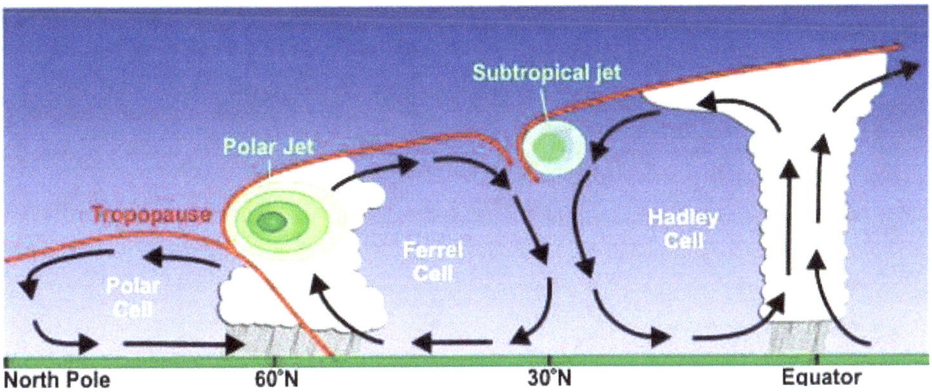

Figure 6.13 Global Circulation Pattern with Jet Streams. Image is used under an Attribution-Share Alike 4.0 International license.

This explains why the polar jet stream is the world's most powerful. Jet streams move seasonally just as the angle of the Sun in the sky migrates north and south. The polar jet stream, known as the jet stream, moves south in the winter and north in the summer.

SUMMARY

The atmosphere is a mixture of nitrogen (78%), oxygen (21%), and other gases (1%) that surround Earth. High above the planet, the atmosphere becomes thinner until it gradually reaches space. It is divided into four layers, based on temperature, and two based on density. Most of the weather and clouds are found in the first layer.

The atmosphere is an important part of what makes Earth livable. It blocks some of the Sun's dangerous rays from reaching Earth. It traps heat, making Earth a comfortable temperature. And the oxygen within our atmosphere is essential for life.

Global and local winds are identified based on which direction the air comes from. Because of rising warm air at specific lines of latitude, warm air can mix with cool air, driving Earth's weather.

7 Chapter | ELEMENTS OF WEATHER AND CLIMATE

Figure 7.1 Mammatus Clouds near Lone Pine, California. Image by Jeremy Patrich is used under a CC-BY 4.0 license.

Goals & Objectives of this unit

- Describe the various aspects and elements of weather and atmospheric water.
- Explain how air masses and weather fronts together form mid-latitude cyclones and
- Describe the three phases a thunderstorm goes through in its life cycle.
- Differentiate between weather and climate and explain their interrelationships.

- Characterize the five general types of climate as defined by the Köppen climate classification.

WEATHER & ATMOSPHERIC MOISTURE

If someone across the country asks you what the weather is like today, you need to consider several factors. Air temperature, humidity, wind speed, the amount and types of clouds, and precipitation are all part of a thorough weather report. In this unit, you will learn about many of these features in more detail. Weather is what is going on in the atmosphere at a particular place at a particular time. Weather can change rapidly. A location's weather depends on air temperature; air pressure; fog; humidity; cloud cover; precipitation; wind speed and direction. All of these are directly related to the amount of energy that is in the system and where that energy is. The ultimate source of this energy is the sun. Climate is the average of a region's weather over time. The climate for a particular place is steady and changes only very slowly.

Climate is determined by many factors, including the angle of the Sun, the likelihood of cloud cover, and the air pressure. All of these factors are related to the amount of energy that is found in that location over time most meteorologists use data spanning nearly 30 years to identify a region's climate.

Humidity

Humidity is the amount of water vapor in the air in a particular parcel of air. We usually use the term to mean relative humidity, the percentage of water vapor a certain volume of air is holding relative to the maximum amount it can contain. If the humidity today is 80%, it means that the air contains 80% of the total amount of water it can hold at that temperature. What will happen if the humidity increases to more than 100%? The excess water condenses and forms precipitation. This is a simplistic look at this topic, because depending on the temperature of the air, the capacity of water content per kilogram of air changes. Warm air can hol d more water vapor than cool air, so raising or lowering the temperature can change the air's relative humidity. The temperature at which air saturated air can condense is called the dew point. This term makes sense, because the water condenses from the air as the dew. A smaller scale example of this would be a cup full of ice water. Depending on the temperature and humidity levels for the day, if the contents in the cup are cooler than the surrounding air, the glass will cause the moisture in the air around the cup to condense along the glass surface.

The image below shows the relationship between relative humidity, dew point and overall air temperature.

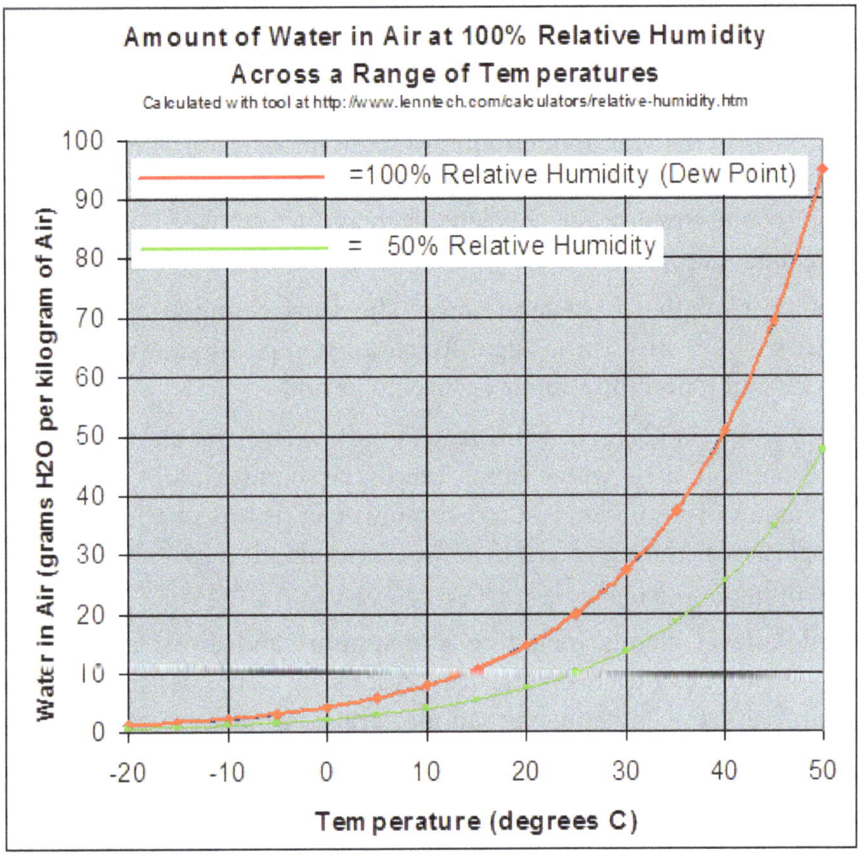

Figure 7.2 Diagram Explaining the Calculation of Relative Humidity. Image is used under a Attribution-Share Alike 3.0 Unported license.

Clouds

Clouds have a big influence on weather by preventing solar radiation from reaching the ground; absorbing warmth that is re-emitted from the ground; and as the source of precipitation. When there are no clouds, there is less insulation. As a result, cloudless days can be extremely hot, and cloudless nights can be very cold. For this reason, cloudy days tend to have a lower range of temperatures than clear days.

There are a variety of conditions needed for clouds to form. First, clouds form when air reaches its dew point. This can happen in two ways:

- Air temperature stays the same but humidity increases. This is common in locations that
- are warm and humid.
- Humidity can remain the same, but temperature decreases. When the air cools enough

- to reach 100% humidity, water droplets form. The air cools when it comes into contact with a cold surface or when it rises.

Rising air creates clouds when it has been warmed at or near the ground level and then is pushed up over a mountain or mountain range or is thrust over a mass of cold, dense air. Water vapor is not visible unless it condenses to become a cloud. Water vapor condenses around a nucleus, such as dust, smoke, or a salt crystal. This forms a tiny liquid droplet. Billions of these water droplets together make a cloud.

Clouds are classified in several ways. The most common classification used today divides clouds into three separate cloud groups which are determined by their altitude and if precipitation is occurring or not.

- High-level clouds form from ice crystals where the air is extremely cold and can hold little water vapor. Cirrus, cirrostratus, and cirrocumulus are all names of high clouds. Cirrocumulus clouds are small, white puffs that ripple across the sky, often in rows. Cirrus clouds may indicate that a storm is coming.
- Middle-level clouds, including altocumulus and altostratus clouds, may be made of water droplets, ice crystals or both, depending on the air temperatures. Thick and broad altostratus clouds are gray or blue-gray. They often cover the entire sky and usually mean a large storm, bearing a lot of precipitation, is coming.
- Low-level clouds are nearly all water droplets. Stratus, stratocumulus, and nimbostratus clouds are common low clouds. Nimbostratus clouds are thick and dark that produce precipitation. Clouds with the prefix 'cumulo-'grow vertically instead of horizontally and have their bases at low altitudes and their tops at high or middle altitudes. Clouds grow vertically when strong unstable air currents are rising upward.

FOG

Fog is a cloud located at or near the ground. When humid air near the ground cools below its dew point, fog is formed. The several types of fog that each form in a different way. Radiation fog forms at night when skies are clear and the relative humidity is high. As the ground cools the bottom layer of air will cool below its dew point. Tule fog is an extreme form of radiation fog found in some regions. San Francisco, California, is famous for its summertime advection fog. Warm, moist Pacific Ocean air blows over the cold California current and cools below its dew point. Sea breezes bring the fog onshore. Steam fog appears in autumn when cool air moves over a warm lake. Water evaporates from the lake surface and condenses as it cools, appearing like steam. Warm humid air travels up a hillside and cools below its dew point to create upslope fog.

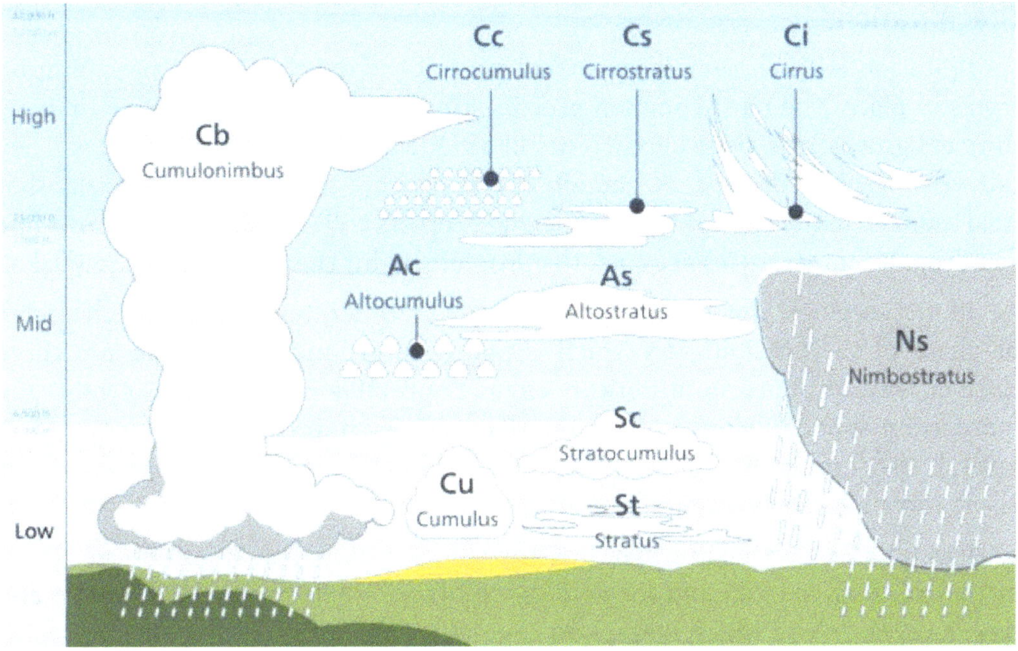

Figure 7.3 Cloud Identification Diagram. Image is used under an Attribution-Share Alike 3.0 Unported license.

- Advection fog: This type of fog forms from surface contact of horizontal winds. This fog
- can occur in windy conditions. Warm air, moist air blows in from the south and if there is snow or cool moisture on the ground it will come in contact with the warm, moist winds. This contact between the air and ground will cause the air blowing in to become cool. Then the dew point rises and creates high humidity and forms fog.
- Radiation fog: This fog forms when all solar energy exits the earth and allows the
- temperature to meet up with the dew point. The best condition to have radiation fog is
- when it had rained the previous night. This helps to moisten up the soil and create higher dew points. This makes it easier for the air to become saturated and form fog. However, the winds must be less than 15 mph to prevent moisture and dry air from mixing.
- Valley Fog: Valley fog forms in the valley when the soil is moist from previous rainfall. As the skies clear solar energy exits earth and allows the temperature to cool near or at the dew point. This form deep fog, so dense it's sometimes called tule fog

Precipitation

Precipitation is an extremely important part of the weather. Some precipitation forms in place. The most common precipitation comes from clouds. Rain or snow droplets grow as they ride air currents in a cloud and collect other droplets. They fall when they become heavy enough to escape from the rising air currents that hold them up in the cloud. Millions of cloud droplets will combine to make only one raindrop. If temperatures are cold, the droplet will hit the ground as a snowflake.

In meteorology, the various types of precipitation often include the character or phase of the precipitation which is falling to ground level. There are three distinct ways that precipitation can occur. Convective precipitation is generally more intense, and of shorter duration, than stratiform precipitation (arranged in layers). Orographic precipitation occurs when moist air is forced upwards over rising terrain, such as a mountain.

Precipitation can fall in either liquid or solid phases, or transition between them at the freezing level. Liquid forms of precipitation include rain and drizzle and dew. Rain or drizzle which freezes on contact within a subfreezing air mass gains the preceding adjective "freezing", becoming known as freezing rain or freezing drizzle. Frozen forms of precipitation include snow, ice crystals, ice pellets (sleet), hail, and graupel. Their respective intensities are classified either by rate of fall or by visibility restriction.

AIR MASSES

Where an air mass receives its characteristics of temperature and humidity is called the source region. Air masses are slowly pushed along by high-level winds when an air mass moves over a new region, it shares its temperature and humidity with that region. The temperature and humidity of a particular location depend partly on the characteristics of the air mass that sits over it. Storms arise if the air mass and the region it moves over have different characteristics. For example, when a colder air mass moves over the warmer ground, the bottom layer of air is heated. That air rises, forming clouds, rain, and thunderstorms. How would a moving air mass form an inversion? When a warmer air mass travels over colder ground, the bottom layer of air cools and, because of its high density, is trapped near the ground.

In general, cold air masses tend to flow toward the equator and warm air masses tend to flow toward the poles. This brings heat to cold areas and cools down warm areas. It is one of the many processes that act towards balancing out the planet's temperatures. Air masses are slowly pushed along by high-level winds. When an air mass moves over a new region, it shares its temperature and humidity with that region. The temperature and humidity of a particular location depend partly on the characteristics of the air mass that sits over it. Air masses are classified based on their temperature and humidity characteristics. Below are examples of how air masses are classified over North America:

- Maritime tropical (mT) – moist, warm air mass
- Continental tropical (cT) – dry, warm air mass
- Maritime polar (mP) – moist, cold air mass
- Continental polar (cP) – dry, cold air mass

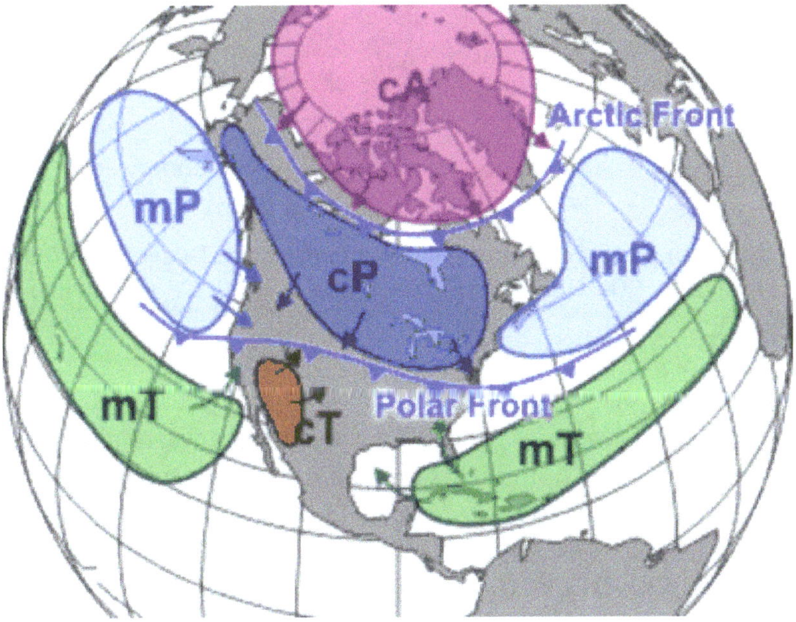

Figure 7.4 Air Masses.Image is in the public domain.

Storms arise if the air mass and the region it moves over have different characteristics. For example, when a colder air mass moves over the warmer ground, the bottom layer of air is heated. That air rises, forming clouds, rain, and sometimes thunderstorms. How would a moving air mass form an inversion? When a warmer air mass travels over colder ground, the bottom layer of air cools and, because of its high density, is trapped near the ground.

In general, cold air masses tend to flow toward the equator and warm air masses tend to flow toward the poles. This brings heat to cold areas and cools down warm areas. It is one of the many processes that act towards balancing out the planet's temperatures.

Weather Front

A front is identified as the zone between two masses of air, and these zones respond differently based on the temperature of the air merging. There are four types of fronts, Cold, Warm, Occluded and Stationary. With cold fronts and warm fronts, the air mass at the leading edge of the front gives the front its name. In other words, a cold front is right at the leading edge of moving cold air and a warm front marks the leading edge of moving warm air.

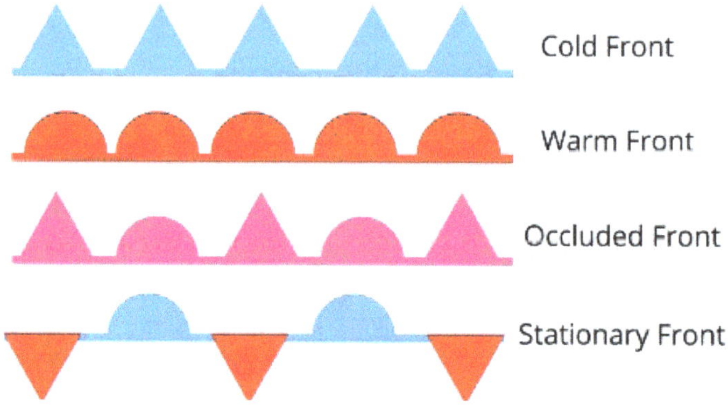

Figure 7.5 Diagram Depicting how Weather Fronts are Drawn on Weather Maps. Image by COC OER Team is used under a CC-BY 4.0 license

COLD FRONT

Imagine that you are standing in one spot as a cold front of air approaches. Along the cold front, the denser, cold air pushes up the warm air, causing the air pressure to decrease. If the humidity is high enough, some types of cumulus clouds will develop. High in the atmosphere, winds blow ice crystals from the tops of these clouds to create cirrostratus and cirrus clouds. At the front, there will be a line of rain showers, snow showers, or thunderstorms with blustery winds. A squall line is a line of severe thunderstorms that forms along a cold front. Behind the front is the cold air mass. This mass is drier, so precipitation stops. The weather may be cold and clear or only partly cloudy. Winds may continue to blow into the low-pressure zone at the front. The weather at a cold front varies with the season.

- Spring & Summer: The air is unstable so thunderstorms or tornadoes may form.
- Spring: If the temperature gradient is high, strong winds blow.
- Autumn: Strong rains fall over a large area.
- Winter: The cold air mass is likely to have formed in the frigid arctic so there are frigid temperatures and heavy snows.

WARM FRONT

Along a warm front, a warm air mass slides over a cold air mass. When warm, less dense air moves over the colder, denser air, the atmosphere is relatively stable. Imagine that you are on the ground in the wintertime under a cold winter air mass with a warm front approaching.

Figure 7.6 A Cold Front (Blue Arrow) Moving in and Forcing an Air Parcel (Green Arrow) Upward.

The transition from cold air to warm air takes place over a long distance so the first signs of changing weather appear long before the front is actually over you. Initially, the air is cold: the cold air mass is above you and the warm air mass is above it. High cirrus clouds mark the transition from one air mass to the other. Over time, cirrus clouds become thicker and cirrostratus clouds form. As the front approaches, altocumulus and altostratus clouds appear and the sky turns gray. Since it is winter, snowflakes fall. The clouds thicken and nimbostratus clouds form. Snowfall increases. Winds grow stronger as the low-pressure approaches. As the front gets closer, the cold air mass is just above you, but the warm air mass is not too far above that. The weather worsens. As the warm air mass approaches, temperatures will rise, and snow turns to sleet and freezing rain. Warm and cold air mix at the front, leading to the formation of stratus clouds and fog.

Figure 7.7 Image of a Warm Front (Green Arrow) Pushing Away a Parcel of Air. .

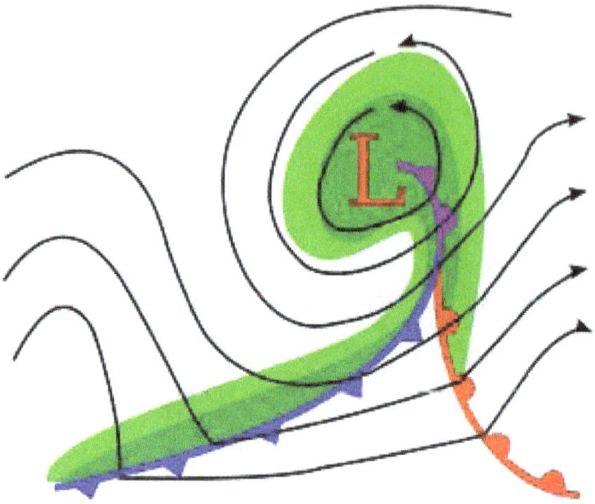

Figure 7.8 Image of an Occluded Front. Image is in the public domain.

OCCLUDED FRONTS

An occluded front usually forms around a low-pressure system. The occlusion starts when a cold front catches up to a warm front. The air masses, in order from front to back, are cold, warm, and then cold again. The Coriolis Effect curves the boundary where the two fronts meet towards the pole. If the air mass that arrives third is colder than either of the first two air masses, that air mass slips beneath them both. This is called a cold occlusion. If the air mass that arrives third is warm, that air mass rides over the other air mass. This is called a warm occlusion. The weather at an occluded front is especially fierce right at the occlusion.

Precipitation and shifting winds are typical. The Pacific Coast has frequently occluded fronts.

Remember, a weather front is the boundary between two air masses of different densities. At the center of each air, mass is typically a high pressure. This means that weather is typicallysunny within air masses, but their temperatures could vary with the season and humidity could vary based on the source region of the air mass.

Now more often than not, these weather fronts are not isolated events. Often, they are part of a larger rotating system called a mid-latitude cyclone. This type of cyclone will be discussed later in this chapter, but as an introduction, it is a low-pressure system that is usually mixing warmer air from the south (in the Northern Hemisphere) and colder air from the north.

STATIONARY FRONT

At a stationary front, the air masses do not move. A front may become stationary if an air mass is stopped by a barrier, such as a mountain range. A stationary front may bring days of rain, drizzle, and fog. Winds usually blow parallel to the front, but in opposite directions. After several days, the front will likely break apart. When a cold air mass takes the place of a warm air mass, there is a cold front.

TYPES OF EXTREME WEATHER

Weather is experienced every day, but only some days experience extreme weather, such as storms. A storm's magnitude can vary immensely depending on whether they are composed of warm or cold air, originating off the ocean or off a continent, occurring in summer or winter, and many other factors. The effects of storms also vary depending on whether they strike a populated area or a natural landscape.

Thunderstorms

Thunderstorms are extremely common: Worldwide there are 14 million per year; that's 40,000 per day. Most precipitate a lot of rain in a small area quickly, but some storms can be severe and highly damaging. Thunderstorms form when ground temperatures are high, ordinarily in the late afternoon or early evening in spring and summer.

All thunderstorms go through a three stage life cycle. The first stage is called the cumulus stage, where an air parcel is forced to rise, cool, and condense, called the lower condensation level, to develop into a cumulus cloud. The process of water vapor condensing into liquid water releases large quantities of latent heat, which makes the air within the cloud warmer, and unstable causing the cloud continues to grow upward like a hot air balloon. These rising air parcels, called updrafts, prevent precipitation from falling from the cloud. But once the precipitation becomes too heavy for the updrafts to hold up, the moisture begins to fall creating downdrafts within the cloud. The downdrafts also begin to pull cold, dry air from outside the cloud towardthe ground in a process called entrainment. Once the precipitation begins to fall from the cloud, the storm has reached the mature stage. During this stage, updrafts and downdrafts exist side- by-side and the cumulonimbus is called a cell. If the updrafts reach the top of the troposphere, the cumulus cloud will begin to spread outward creating a defined anvil. At the same time, the downdrafts spread within the cloud and at first make the cloud become wider, but eventually overtaking the updrafts. Cool downdrafts form when precipitation and the cool airfrom entrainment are dragged down to the lower regions of a thunderstorm. It is also during the mature stage when the storm is most intense producing strong, gusting winds, heavy precipitation, lightning, and possibly hail.

Once the downdrafts overtake the updrafts, which also prevents the release of latent heat energy, the thunderstorm will begin to weaken into the third and final stage, calledthe dissipating stage. During this stage, light precipitation and downdrafts become the dominant feature within the cloud as it weakens. In all, only 20% of the moisture within the cloud fell as precipitation whereas the other 80% evaporates back into the atmosphere.

With severe thunderstorms, the downdrafts are so intense that when they hit the ground it sends warm air from the ground upward into the storm. The warm air gives the convection cells more energy. Rain and hail grow before gravity pulls them to Earth. Severe thunderstorms can last for hours and can cause a lot of damage because of high winds, flooding, intense hail, and tornadoes. Thunderstorms can form individually or in squall lines along a cold front. In the United States, squall lines form in spring and early summer in the Midwest where the maritime tropical (mT) air mass from the Gulf of Mexico meets the continental polar (cP) air mass from Canada.

So much energy collects in cumulonimbus clouds that a huge release of electricity,called lightning, may result. The electrical discharge may be between one part of the cloud and another, two clouds, or a cloud and the ground.

Tornadoes

Tornadoes, also called twisters, are fierce products of severe thunderstorms. As the air in a thunderstorm rises, the surrounding air races in to fill the gap, forming a funnel. A tornado lasts from a few seconds to several hours. The average wind speed is about 177 kph (110 mph), but some winds are much faster. A tornado travels over the ground at about 45 km per hour (28 miles per hour) and goes about 25 km (16 miles) before losing energy and disappearing.

An individual tornado strikes a small area, but it can destroy everything in its path. Most injuries and deaths from tornadoes are caused by flying debris. In the United States, an average of 90 people is killed by tornadoes each year. The most violent 2% of tornadoes account for 70% ofthe deaths by tornadoes. Tornadoes form at the front of severe thunderstorms. Lines of these thunderstorms form in the spring where maritime tropical (mT) and continental polar (cP) air masses meet. Although there is an average of 770 tornadoes annually, the number of tornadoes each year varies greatly.

In late April 2011, the mid-west region of the United States experienced a tornado Super Outbreak, totaling over 300 tornadoes, traveling through 15 states, in only three days. In addition to the meeting of cP and mT, the jet stream was blowing strongly in from the west. The entire region was alerted to the possibility of tornadoes in those late April days. But meteorologists can only predict tornado danger over a very wide region. No one can tell exactly where and when a tornado will touch down. Once a tornado is sighted on the radar, its path is predicted, and a warning is issued to people in that area. The exact path is unknown because the tornado movement is not very predictable.

The intensity of tornadoes is measured on the Fujita Scale, which assigns a value based on wind speed and damage.

Enhanced Fujita Scale

EF0	65–85 mph	Light damage
EF1	86–110 mph	Moderate damage
EF2	111–135 mph	Considerable damage
EF3	136–165 mph	Severe damage
EF4	166–200 mph	Devastating damage
EF5	>200 mph	Incredible damage

Figure 7.9 The Enhanced Fujita Scale. (Image on Wikimedia Commons by Pfly CC BY-SA 3.0)

Cyclones

Cyclones can be the most intense storms on Earth. A cyclone is a system of winds rotating counterclockwise in the Northern Hemisphere around a low-pressure center. The swirling air rises and cools, creating clouds and precipitation.

There are two types of cyclones: middle latitude (mid-latitude) cyclones and tropical cyclones. Mid-latitude cyclones are the main cause of winter storms in the middle latitudes. Tropical cyclones are also known as hurricanes.

Mid-Latitude Cyclones

Mid-latitude cyclones, sometimes called extratropical cyclones, form at the polar front when the temperature difference between two air masses is large. These air masses blow past each other in opposite directions. Coriolis Effect deflects winds to the right in the Northern Hemisphere, causing the winds to strike the polar front at an angle. Warm and cold fronts form next to each other. Most winter storms in the middle latitudes, including most of the United States and Europe, are caused by mid-latitude cyclones. The warm air at the cold front rises and creates a low-pressure cell. Winds rush into the low pressure and create a rising column of air. The air twists, rotating counterclockwise in the Northern Hemisphere and clockwise in the Southern Hemisphere. If the rising air contains enough moisture, rain, or snow may fall.

Mid-latitude cyclones form in winter in the mid-latitudes and move eastward with the westerly winds. These two- to five-day storms can reach 1,000 to 2,500 km (625 to 1,600 miles) in diameter and produce winds up to 125 km (75 miles) per hour. Like tropical cyclones, they can cause extensive beach erosion and flooding.

Mid-latitude cyclones are especially fierce in the mid-Atlantic and New England states where they are called nor'easters, because they come from the northeast. About 30 nor'easters strike the region each year.

Hurricanes

Tropical cyclones have many names. They are called hurricanes in the North Atlantic and Eastern Pacific oceans, typhoons in the western Pacific Ocean, tropical cyclones in the Indian Ocean, and willi-willi's in the waters near Australia. By any name, they are the most damaging storms on Earth. Hurricanes arise in the tropical latitudes (between 10° and 25°N) in summer and autumn when sea surface temperatures are 28° C (82° F) or higher. The warm seas create a large humid air mass. The warm air rises and forms a low-pressure cell, known as a tropical depression. Thunderstorms materialize around the tropical depression.

If the temperature reaches or exceeds 28° C (82° F) the air begins to rotate around the low pressure (counterclockwise in the Northern Hemisphere and clockwise in the Southern Hemisphere). As the air rises, water vapor condenses, releasing energy from latent heat. If wind shear is low, the storm builds into a hurricane within two to three days.

Hurricanes are large systems with high winds. The exception is the relatively calm eye of the storm where the air is rising upward. Rainfall can be as high as 2.5 cm (1″) per hour, resulting in about 20 billion metric tons of water released daily in a hurricane. The release of latent heat generates enormous amounts of energy, nearly the total annual electrical power consumptionof the United States from one storm. Hurricanes can also generate tornadoes. Hurricanes are strange phenomena because they are deadly monsters, yet have a gentle, but coldheart. The anatomy of a hurricane is fairly simple, though the processes involved are quite complex. As a low-pressure disturbance forms, the warm, moist air rushes towards the low pressure to rise upward to form towering thunderstorms. Around the low-pressure disturbance is a wall of clouds called an eyewall. Within the eyewall, the wind speeds are the greatest, the clouds are the tallest, the atmospheric pressure is at its lowest, and precipitation is most intense.

Moving away from the eyewall are organized, intense thunderstorms, called spiral rain bands that rotate around and toward the storm's eyewall. These rainbands are the first hurricanes are assigned to categories based on their wind speed. The categories are listed on the Saffir- Simpson Scale.

Hurricanes move with the prevailing winds. In the Northern Hemisphere, they originate in the trade winds and move to the west. When they reach the latitude of the westerlies, they switch direction and travel toward the north or northeast. Hurricanes may cover 800 km (500 mi) in one day.

Table 7.1 Saffir – Simpson Hurricane Scale

Category	Wind Speed (mph)	Type of Damage
1	74 – 95	Some Damage
2	96 – 110	Extensive Damage
3	111 – 129	Devastating
4	130 – 156	Catastrophic Damage
5	157 and above	Catastrophic Damage

Damage from hurricanes comes from the high winds, rainfall, and storm surge. Storm surge occurs as the storm's low-pressure center comes onto land, causing the sea level to rise unusually high. A storm surge is often made worse by the hurricane's high winds blowing seawater across the ocean onto the shoreline. Flooding can be devastating, especially along low-lying coastlines such as the Atlantic and Gulf Coasts. Hurricane Michael in 2018 had peak winds of 260 km/h (160 mph) and storm surges up to 4.3 m (14 ft).

Hurricanes typically last for 5 to 10 days. Over cooler water or land, the hurricane's latent heat source shutdowns and the storm weakens. When a hurricane disintegrates, it is often replaced with intense rains and tornadoes.

WEATHER VERSUS CLIMATE

People often confuse weather and climate; they are not identical. According to the American Meteorological Society (AMS), the weather is defined as the state of the atmosphere at some place and time, usually expressed in terms of temperature, air pressure, humidity, wind speed and direction, precipitation, and cloudiness. Meteorologists study the atmosphere, processes that cause weather, and the life cycle of weather systems.

Table 7.2 Basic Characteristics of the Köppen Climate Classification

Type of Climate	Characteristics
Tropical (A)	Humid average temperate above 18° C (64°F)
Dry (B)	Evaporation exceeding precipitation with constant water deficiency throughout the year
Temperate (C)	Humid and warm or hot summers and mild winters with average temperatures between -3°C (27°F) and 18°C (64°F)
Continental (D)	Humid and warm summers with the average temperature of warmest
Polar (E)	Extremely cold winters and an average temperature of the warmest summer month below 10°C (50°F)

Climate is defined in terms of the average (mean) of weather elements (such as temperature and precipitation) over a specified period of time. (The World Meteorological Organization defines the typical time period of time as 30 years). Climate also encompasses weather extremes for a particular place.

Scientists have developed a variety of ways for classifying climate. In the early 20th century, a German scientist named Vladimir Köppen developed one of the most widely used classification systems. The Köppen system categorizes climate into five main types, which can be further divided into subcategories.

The planet's climate has changed many times over Earth's long geologic history. For example, over the past million years, Earth has experienced several glacial periods interspersed with interglacial (warmer) periods. The relatively constant and favorable interglacial period of climate experienced over the past 8,000 years has made human civilization's advancement possible.

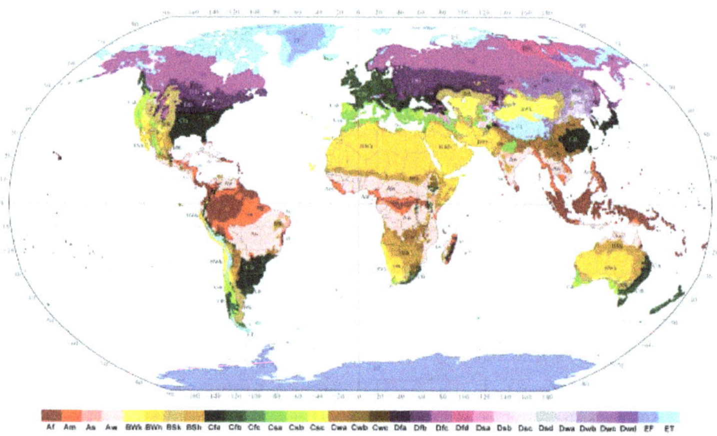

Figure 7.10 Köppen-Geiger Climate Subdivisions. Image is in the public domain.

Climate Change

Climate change refers to a significant and sustained (over decades or longer) change from one climatic condition to another. The term "global warming" refers to a specific kind of climate change in which Earth's average temperature is increasing. Of growing concern is what is known as abrupt climate change. According to the National Oceanic and Atmospheric Administration (NOAA), abrupt climate change is a relatively new area of scientific research whose formal definition is still being developed, but it refers to a sudden, rapid change from one climate state to another (over a period of years rather than centuries or millennia).

Meteorologists focus primarily on real-time (current) data to predict local or regional atmospheric conditions for the hours, days, or weeks ahead. Thus, weather prediction tends to be more local and relates to conditions in the immediate future from days to weeks.

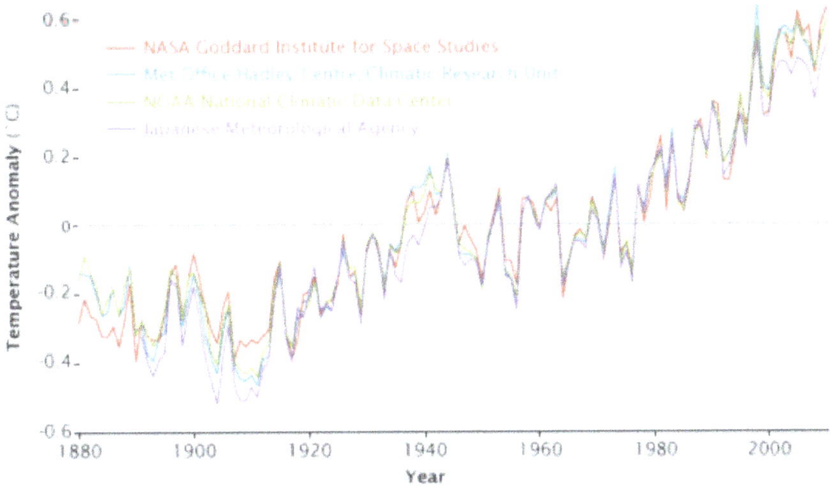

Figure 7.11 Annual Temperatures During 1880-2018. Notice the Winter/Summer Patterns—& Then the Overall Dramatic Increase of The Patterns During the Past 20 Years. Image is in the public domain.

Climate scientists or climatologists, on the other hand, look at atmospheric conditions in terms of averages and trends (patterns) that have occurred over many decades, centuries, and millennia. Weather is variable but can be averaged over time to indicate climate trends.Therefore, climate scientists can use weather data plus proxy data to help them identify previous trends and improve their predictions of future trends.

Meteorologists and climate scientists use similar tools. Weather balloons, satellites, specially designed airplanes, and radar and other ground-based data collection instruments (to measure wind speed, precipitation, air temperature, humidity levels, etc.) are all good examples. These methods and tools have enabled humans to collect reliable atmospheric data on a consistent basis since the mid-1800s. They have grown increasingly more precise and sophisticated over time, to such an extent that meteorologists can now consistently provide reasonably accurate near-term (1 week or less) weather forecasts.

Climate monitoring requires data covering all areas of the planet over a much longer time period. Sophisticated Earth-observing satellites equipped with remote-sensing equipment circle the globe. With each pass, they can record sea surface and other temperatures, measure atmospheric gases and rainfall amounts, take visible and infrared photos of Earth's surface and calculate Earth's outgoing infrared and reflected solar radiation.

Extreme Weather

All-weather events that cause loss of life, disrupt normal human activities, and result in property damage appear extreme. It is a question of perspective: How do today's severe weather events compare to severe weather events in the recent and distant past? The resolution of Global Climate Models can complicate making direct comparisons between past and present events. For example, since 1986 the global human population has grown by approximately 2 billion. Simply said, there are more people than ever living in formerly unpopulated or sparsely populated areas. Comparing death tolls, between recent and past events may not be the most meaningful indicator of a particular weather event's intensity.

Nonetheless, the growing body of meteorological data indicates an increase in the number of extreme weather events occurring here in the United States since 1980, and the number of extreme events also appears to be rising worldwide.

SUMMARY

Weather, the state of the atmosphere at a particular place during a short period of time. It involves atmospheric phenomena such as temperature, humidity, precipitation (type and amount), air pressure, wind, and cloud cover.

An air mass is a large mass of air that has similar characteristics of temperature and humidity within it. An air mass acquires these characteristics above an area of land or water known as its source region. When the air mass sits over a region for several days or longer, it picks up the distinct temperature and humidity characteristics of that region. There are four types of air masses.

A weather front is a transition zone between two different air masses at the Earth's surface. Each air mass has a unique temperature and humidity characteristics. Often there is turbulence at a front, which is the borderline where two different air masses come together. The turbulence can cause clouds and storms.

The Köppen climate classification is one of many systems that help identify a region's climate. There are five main climate groups, with each group being divided based on seasonal precipitation and temperature patterns.

The weather differs from the climate in that the latter includes the synthesis of weather conditions that have prevailed over a given area during a long time period—generally 30 years.

Climate, by contrast, refers to weather trends and patterns occurring globally or regionally over decades, centuries, and even millennia. Extreme weather events, by definition, are rare and intense. You have learned that although scientists are still unable to conclusively link specific extreme weather events to global climate change, these events are predicted consequences of long-term changes in Earth's climate.

8

Chapter

BASIC MINERAL DEVELOPMENT

Figure 8.1 Halite Crystals. Image by Jeremy Patrich is used under a CC-BY 4.0 license.

Goals & Objectives of this unit

- Explain how minerals are formed and identified.
- Describe how color, luster, and streak are used to identify minerals.
- Identify additional properties that can be used to identify some minerals.

MATERIALS OF EARTH'S CRUST

The best way to learn about Earth's crust would be to travel around the world, viewing minerals, rocks, and structures in a variety of places to see what they are and how they can be coaxed into telling Earth's story.

All matter is made of tiny particles. Protons, neutrons, and electrons form atoms that bond together to create molecules. Atoms are the smallest units that have the properties of the element they are, and molecules are the smallest units of a compound. For example, water is made of hydrogen and oxygen, but a molecule of water is very different from an atom of hydrogen or an atom of oxygen. The atoms combine to form molecules by different types of chemical bonding. Molecules bond into structures as well. The structures created by molecules form the different types of minerals, most importantly silicates, which are the substances that make up most of Earth's crust. Other important minerals are carbonates and native elements, which are some of the most important materials used by society. Minerals come together to create the three major rock types, igneous, sedimentary, and metamorphic. These rocks are the material part of the rock cycle, (which will be discussed in the following units).

Atoms to Molecules

Everything you can see, touch, smell, feel, and taste is made of atoms. Atoms are the basic building block of all matter, so if we want to know about what Earth is made of, then we have to know a few things about these incredibly small objects.

Everyday experience should convince you that matter is found in myriad forms, yet all the matter you have ever seen is made of atoms, or atoms stuck together in configurations of dizzying complexity. A chemical element is a substance that cannot be made into a simpler form by ordinary chemical means. The smallest unit of a chemical element is an atom, and all atoms of a particular element are identical.

There are two parts to an atom. At the center of an atom is a nucleus made up of two types of particles called protons and neutrons.

- Protons have a positive electrical charge. The number of protons in the nucleus determines what element the atom is.
- Neutrons are about the size of protons but have no charge.

Electrons, much smaller than protons or neutrons, have a negative electrical charge, move at nearly the speed of light, and orbit the nucleus at exact distances, depending on their energy.

IONS

Atoms are stable when they have a full outermost electron energy level. To fill its outermost shell, an atom will give, take, or share electrons. When an atom either gains or loses electrons, this creates an ion. Ions have either a positive or a negative electrical charge. What is the charge of an ion if the atom loses an electron? An atom with the same number of protons and electrons has no overall charge, so if an atom loses the negatively charged electron, it will have more protons, therefore, a positive charge. Ions with a positive charge are referred to as a cation (pronounced Cat-Ion). What is the charge of an ion if the atom gains an electron? If the atom gains an electron, there will be more electrons and will have a negative charge. Anions are ions that have a negative charge.

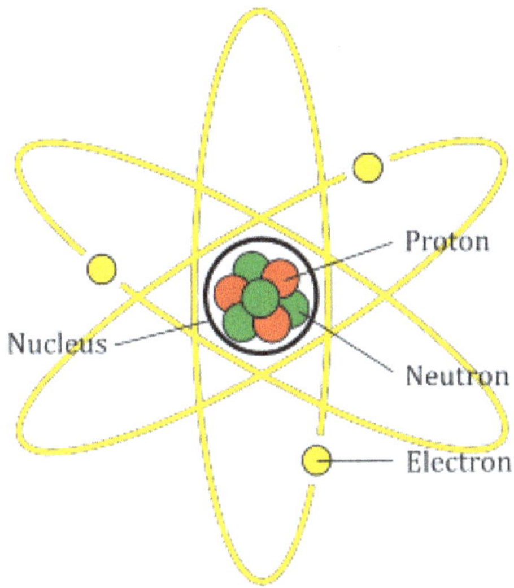

Figure 8.2 The Anatomy of an Atom. Image is used under a Attribution-Share Alike 4.0 International license.

MOLECULES

When a cation gets close to an anion, they link up because of their different net charges, positive charges attract negative charges, and vice versa. When two or more atoms link up, they create a molecule. A molecule of water is made of two atoms of hydrogen (H) and one atom of oxygen (O). The molecular mass is the sum of the masses of all the atoms in the molecule. A collection of molecules is called a compound.

BASIC MINERAL IDENTIFICATION

Minerals can be identified by their physical characteristics. The physical properties of minerals are related to their chemical composition and bonding. Some characteristics, such as amineral's hardness, are more useful for mineral identification. Color is readily observable and certainly obvious, but it is usually less reliable than other physical properties.

Check out the mineral in Figure 8.3 below. What is the mineral's color, & shape? Are the individual crystals shiny or dull? Is there a definite crystalline structure of these minerals? In this unit, the properties used to identify minerals are described in more detail.

Figure 8.3 Fluorite Crystals. Image by Jeremy Patrich is used under a CC-BY 4.0 license.

Color

Color is the most eye-catching feature of many minerals. Some minerals will always have a similar color, such as Gold, whereas some minerals, such as Quartz and Calcite, come in a variety of colors. The presence and intensity of certain elements will determine a specific mineral's color. Minerals with an inherent color (e.g. all specimens of the mineral are the same color) have essential elements in them which cause their color. Good examples are Azurite and Malachite, which has their strong blue and green color due to their copper in their atomic

structure. But there are many minerals which have slight additions of color-causing elements in some specimens that cause it to be a different color. For example, pure Quartz (SiO2), is colorless, whereas Amethyst, a purple variety of quartz, has its purple color caused by traces of the element iron. The amount of iron present determines the intensity of the color.

Certain minerals exhibit a color change when exposed to light, heat, radiation, orwhen atomic anomalies are present. Red Realgar transforms into yellow Paraealgar uponrepeated exposure to light. Some minerals, such as Proustite and Vivianite, darken upon prolonged exposure to light, whereas other minerals, such as Kunzite fade. Some minerals undergo color changes when put under intense heat. This method is commonly used in the gemstone industry to artificially enhance the color of many gemstones. For example, some varieties of Topaz, Beryl, and Corundum are heat-treated to produce deep colored gemstones from duller stones. Radioactivity can also have an effect, as is the cause of the color of Smoky Quartz.

Figure 8.4 Examples of Fluorescent Mineral, Reacting Under Ultraviolet Light (Image on Wikimedia Commons by Pfly, CC BY-SA 3.0)

Streak

Streak is the color of a mineral's powder. Streak is a more reliable property than color becausestreak does not vary. Minerals that are the same color may have a different colored streak. Many minerals, such as the rose quartz in the figure above, do not have astreak, as they are harder than the plate.

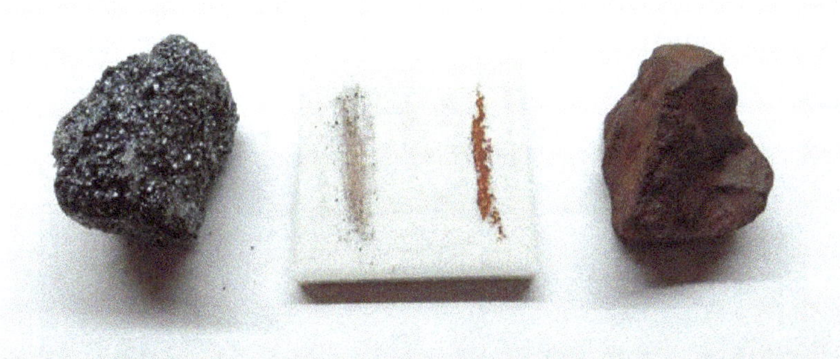

Figure 8.5 Hematite Leaves a Reddish-Brown Streak on a Ceramic Plate Whether the Hematite has a Metallic Luster or an Earthy Luster. Image by is licensed under a Attribution-Share Alike 4.0 International license.

o check streak, scrape the mineral across an unglazed porcelain plate (Figure below). Yellow- gold pyrite has a blackish streak, another indicator that pyrite is not gold, which has a golden yellow streak. The streak of hematite across an unglazed porcelain plate is red-brown.

The streak test should be done on clean, weathered, or freshly broken specimens of the mineral. This is done to reduce the possibility that a contaminant, weathered coating, or tarnish will influence the results of the test. Select a representative point or protrusion on the specimen that will be scraped across the streak plate.

With your other hand, place the streak plate flat on a tabletop or laboratory bench. Then, while holding the streak plate flat and firmly in place on the tabletop, place the point of the specimen firmly against the streak plate, and, while maintaining firm pressure, drag the specimen across the plate.

Luster

- Luster describes the reflection of light off a mineral's surface. Mineralogists have special terms to describe luster. One simple way to classify luster is based on whether the mineral is metallic or non-metallic. Minerals that are opaque and shiny, such as pyrite, have a metallic luster.
- Minerals such as quartz have a non-metallic luster.

Figure 8.6 Examples of Luster—(a) Diamond Has an Adamantine Luster. (b) Quartz is Not Sparkly & has a Vitreous/Glassy Luster. (b) Sulfur Reflects Less Light, so it has a Resinous Luster. Image is used under a CC BY: Attribution license.

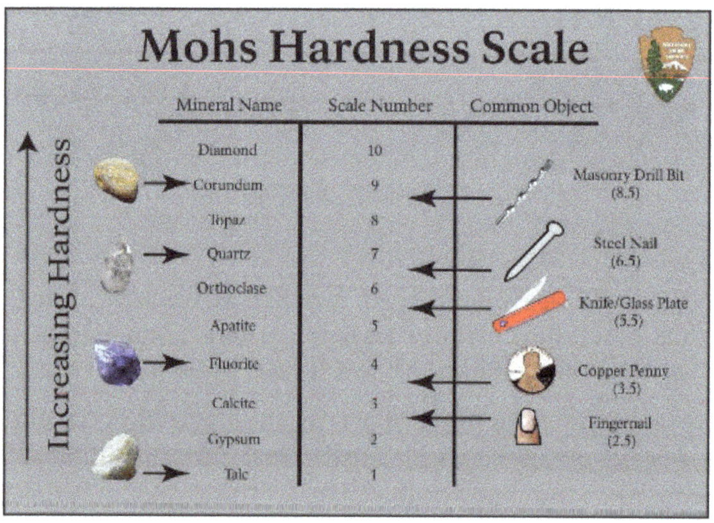

Figure 8.7 Mohs Hardness Scale. Image is in the public domain.

Hardness

The Mohs' scale of mineral hardness is named after Friedrich Mohs, a mineralogist who invented a scale of hardness based on the ability of one mineral to scratch another. Rocks are made up of one or more minerals. According to the scale, Talc is the softest: it can be scratched by all other materials. Gypsum is harder as it can scratch talc but not calcite, which is even harder.

The hardness of a mineral is mainly controlled by the strength of the bonding between the atoms and partly by the size of the atoms. It is a measure of the resistance of the mineral to scratching, the Mohs scale is for natural minerals.

Diamond is always at the top of the scale, being the hardest mineral. There are ten minerals in Mohs scale, talc, gypsum, calcite, fluorite, apatite, feldspar, quartz, topaz, corundum, and for last and hardest, diamond. Because the Mohs scale was made long ago, it is not exactly correct, for example, several minerals are now known to be harder than the diamond. The Mohs scale may not be perfect, but field geologists still find it very useful.

CLEAVAGE & FRACTURE

Cleavage refers to the way some minerals break along certain lines of weakness in their structure. Mica is a good example as it breaks along very closely spaced flat planes that yieldthin sheets. Calcite is another good example, breaking along three different planes that yield blocky fragments that look like a rectangular box that has been warped, called a rhombohedron, simply, "rhomb." Galena breaks along three planes at right angles to one another, producing true cubes as fragments.

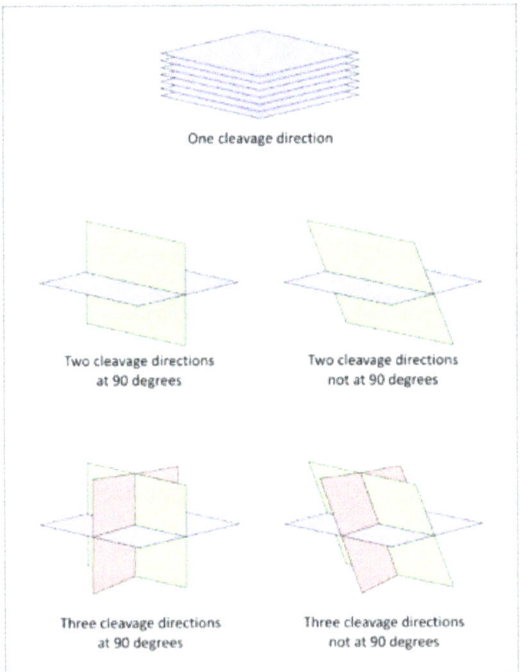

Figure 8.8 Examples of Mineral Cleavage. Image is used under a Creative Commons Attribution-Share Alike 3.0 Unported license.

Cleavages are described in terms of their quality and how smoothly the mineral breaks, their resistance to fracture, and resulting shape. The quality of cleavages is perfect, imperfect, distinct, good, fair, and poor. The difficulty is described as easy, hard, and difficult to produce. By way of examples, the micas have perfect cleavage in one direction that is easy to produce; calcite has a perfect cleavage in three directions that is also easy to produce; the feldspars have perfect cleavage in one direction that is easy to produce and a good cleavage in another direction that is hard to produce; and diamond has a perfect cleavage in four directions that is easy to produce. Sphalerite has perfect cleavages in six directions, some of which are easy to produce, others hard, hence you won't always see all six cleavage surfaces in any given sample of the material.

Cleavage is the tendency of a mineral to break along certain planes to make smooth surfaces. Halite breaks between layers of sodium and chlorine to form cubes with smooth surfaces.

Figure 8.9 Mica Exhibits One-Plane of Cleavage—Sheets. Image is used under a CC BY: Attribution license.

Fracture is a break in a mineral that is not along a cleavage plane. Fracture is not always the same in the same mineral because fracture is not determined by the structure of the mineral. Minerals may have characteristic fractures. Metals usually fracture into jagged edges. If the mineral splinters like wood, it may be fibrous. Some minerals, such as quartz, form smooth curved surfaces when they fracture.

Figure 8.10 Chrysotile, Exhibits a Splintery Fracture. Image is used under a CC BY: Attribution license.

SUMMARY

An atom has negatively charged electrons in orbit around its nucleus, which is composed of positively-charged protons and neutrons, which have no charge.

An atom that gains or loses electrons is an ion. Positively charged ions are cations, negatively charged ions are anions.

Minerals can be identified by their physical characteristics. The physical properties of minerals are related to their chemical composition and bonding.

Some characteristics, such as a mineral's hardness, are more useful for mineral identification. Color is readily observable and certainly obvious, but it is usually less reliable than other physical properties.

The International Mineralogical Association has established the following requirements for a substance to be considered a distinct mineral:

It must be a naturally occurring substance formed by natural geological processes, on Earth or other extraterrestrial bodies. This excludes compounds directly and exclusively generated by human activities (anthropogenic) or in living beings (biogenic), such as tungsten carbide, urinary calculi, calcium oxalate crystals in plant tissues, and seashells.

It must be a solid substance in its natural occurrence.

It must have a well-defined crystallographic structure; or, more generally, an ordered atomic arrangement.

It must have a fairly well-defined chemical composition. However, certain crystalline substanceswith a fixed structure but variable composition may be considered single mineral species.

9 Chapter | IGNEOUS ROCKS

Figure 9.1 Lake Sabrina, a Glacially Carved Lake in the Eastern Sierra. Image by Jeremy Patrich is used under a CC-BY 4.0 license.

Goals & Objectives of this unit

- Describe the rock cycle and the types of processes that lead to the formation of igneous,
- sedimentary, and metamorphic rocks, and explain why there is an active rock cycle on
- Earth.
- Describe, in general terms, the range of chemical compositions of magmas.
- Discuss the processes that take place during the cooling and crystallization of magma, and the typical order of crystallization according to the Bowen reaction series.

- Explain how magma composition can be changed by fractional crystallization and partial
- melting of the surrounding rocks.
- Apply the criteria for igneous rock classification based on mineral proportions.
- Describe the origins of phaneritic, porphyritic, and pegmatitic textures.

THE ROCK CYCLE

The rock components of the crust are slowly but constantly being changed from one form to another and the processes involved are summarized in the rock cycle. The rock cycle is driven by two forces:

- Earth's internal heat engine, which moves material around in the core and the mantleand leads to slow but significant changes within the crust.
- The hydrological cycle, which is the movement of water, ice, and air at the surface, and is powered by the sun.

The rock cycle is still active on Earth because our core is hot enough to keep the mantle moving, our atmosphere is relatively thick, and we have liquid water. On some other planets or their satellites, such as the Moon, the rock cycle is virtually dead because the core is no longer hot enough to drive mantle convection and there is no atmosphere or liquid water.

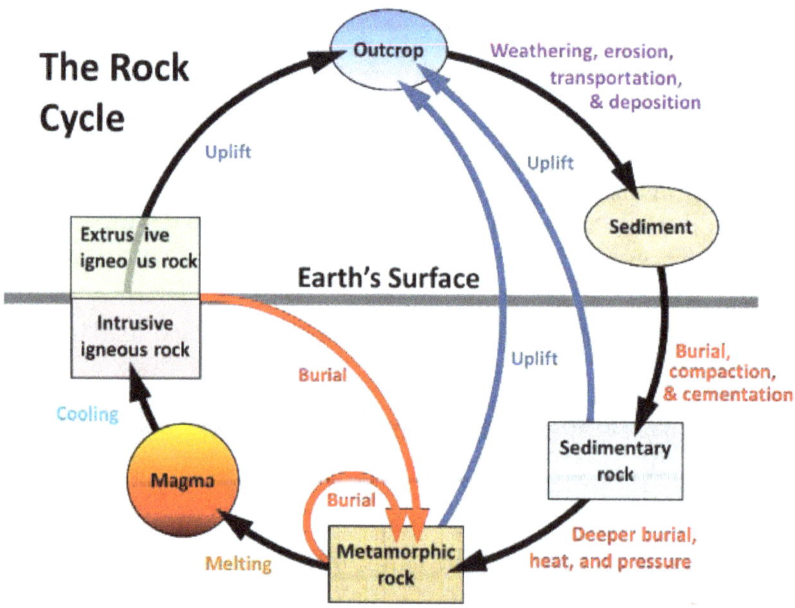

Figure 9.2 The Rock Cycle. Image is used under a Creative Commons Attribution 4.0 International License.

In describing the rock cycle, we can start anywhere we like, although it's convenient to start with magma. Magma is a rock that is hot to the point of being entirely molten. This happens at between about 800° and 1300°C, depending on the composition and the pressure, onto the surface, and cool quickly (within seconds to years), forming extrusive igneous rock.

Magma can either cool slowly within the crust (over centuries to millions of years), forming an intrusive igneous rock or can erupt onto the surface and cool quickly (within seconds to years) forming an extrusive igneous rock. An intrusive igneous rock typically crystallizes at depths of hundreds of meters to tens of kilometers below the surface. To change its position in the rock cycle, intrusive igneous rock has to be uplifted and exposed by the erosion of the overlying rocks.

Through the various plate-tectonics-related processes of mountain building, all types of rocks are uplifted and exposed at the surface. Once exposed, they are weathered, both physically (by the mechanical breaking of the rock) and chemically (by weathering of the minerals), and the weathering products, mostly small rock and mineral fragments are eroded, transported, and then deposited as sediments. Transportation and deposition occur through the action of glaciers, streams, waves, wind, and other agents, and sediments are deposited in rivers, lakes, deserts, and the ocean.

Figure 9.3 Collecting Diorite (Granite) Samples from a Fresh Rockslide near Lone Pine CA. Image by Jeremy Patrich is used under a CC-BY 4.0 license.

Granite is nearly always massive, hard, and tough. These properties have made granite a widespread construction stone throughout human history. The term "granitic" means granite- like and is applied to granite and a group of intrusive igneous rocks with similar textures and slight variations in composition and origin. In most cases, a body of hot magma is less dense than the rock surrounding it, so it tendsto move very slowly up toward the surface. It does so in a few different ways, including filling and widening existing cracks, melting the surrounding rock (called country rock), pushing the rock aside (where it is somewhat plastic), and breaking the rock. Where some of the country rock is broken off, it may fall into the magma, a processcalled stoping. The resulting fragments are known as xenoliths (Greek for "strange rocks").

Figure 9.4 A Xenolith in Granite, Near Rock Creek (Eastern Sierra, California). (Image on Wikimedia Commons by Pfly, CC BY-SA 3.0)

Figure 9.5 Half Dome in California is an Example of Masses of Rocks, or Plutons, Creating a Batholith. Image by DAVID ILIFF is used under a CC BY-SA 3.0 license.

Some upward-moving magma reaches the surface, resulting in volcanic eruptions, but most cools within the crust. The resulting body of rock is known as a pluton.

Large irregular-shaped plutons are called either stocks or batholiths. The distinction between the two is based on the area that is exposed at the surface: if the body has an exposed surface area greater than 100 km2, then it's a batholith; smaller than 100 km2 and it's a stock.

Batholiths are typically formed only when a number of stocks coalesce beneath the surface to create one large body. One of the largest batholiths in the world is the Coast Range Plutonic Complex, which extends from the Vancouver region in Canada, to southeastern Alaska. More accurately, it's many batholiths.

Tabular (sheet-like) plutons are distinguished based on whether or not they are concordant with (parallel to) existing layering (e.g., sedimentary bedding or metamorphic foliation) in the country-rock. A sill is concordant with existing layering, and dyke is discordant. If the country rock has no bedding or foliation, then any tabular body within it is a dyke. Note that the sill-versus-dyke designation is not determined simply by the orientation of the feature. A dyke can be horizontal, and a sill can be vertical (if the bedding is vertical).

TEXTURES

Igneous rocks are also classified according to their textures. The textures of volcanic rocks will be discussed later, so here we'll only look at the different textures of intrusive igneous rocks. Almost all intrusive igneous rocks have crystals that are large enough to see with the naked eye, and we use the term phaneritic (from the Greek word phaneros meaning visible) to describe that. Typically, that means they are larger than about 0.5 mm, or the thickness of a strong line made with a ballpoint pen. If the crystals are too small to distinguish, which is typical of most volcanic rocks, we use the term aphanitic.

In general, the size of the crystals is proportional to the rate of cooling. The longer it takes for a body of magma to cool, the larger the crystals will be. It is not uncommon to see an intrusive igneous rock with crystals up to a centimeter long. In some situations, especially toward the end of the cooling stage, the magma can become water-rich. The presence of liquid water (still liquid at high temperatures because it is under pressure) promotes the relatively easy movement of ions, and this allows crystals to grow large, sometimes to several centimeters. As already described, if an igneous rock goes through a two-stage cooling process, its texture will be porphyritic, as seen below in the close-up of a granite sample.

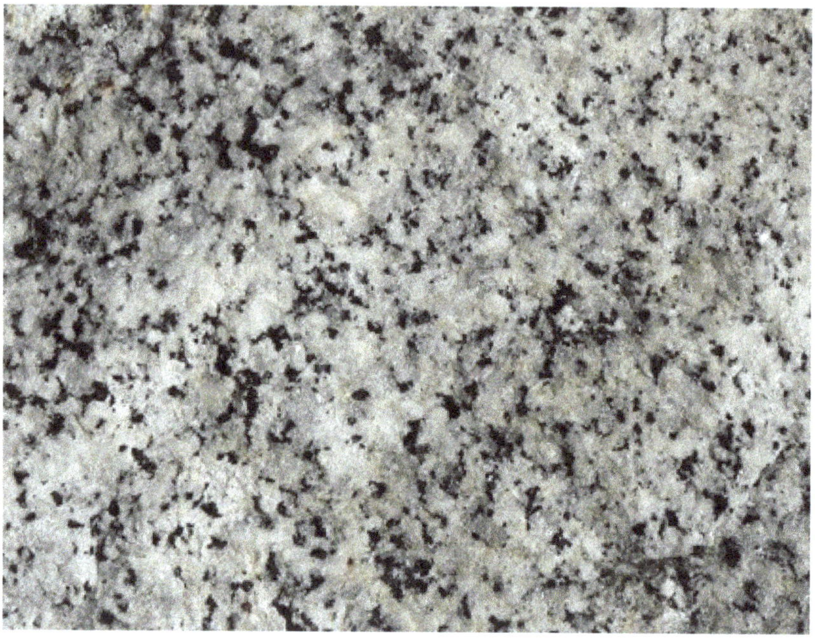

Figure 9.6 Close-Up of a Granite Sample. This Sample, it is Easy to See & Identify the White & Black Crystals, so This Specimen Cooled Slowly. Jeremy Patrich is used under a CC-BY 4.0 license.

Igneous rocks are called extrusive or volcanic when they form above the surface. They solidify after molten material pours out onto the surface through an opening such as a volcano.

Figure 9.7 Close-Up of an Obsidian Sample. This Extrusive Rock Cooled Quickly from Felsic Lava. Jeremy Patrich is used under a CC-BY 4.0 license.

MAGMA

Magma is a complex high-temperature fluid substance. Temperatures of most magmas are in the range of 700°C to 1300°C. Magma can get forced into adjacent rocks (intrusion or plutonic) or forced out to the surface (extrusion or volcanic) as lava or blown out in explosions which include rock pieces (tephra). Magma is made up of atoms and molecules of melted minerals. When magma cools the atoms and molecules rearrange to form mineral grains. Rock forms when mineral grains (often crystals) grow together. Granite, diorite, gabbro, and basalt are a few types of igneous rock. Quartz is one of the chief minerals produced by igneous action; it is made of silica (SiO2), the most common molecule in igneous minerals. Some examples of igneous volcanic rock are pumice, obsidian (volcanic glass), and scoria, and much more. Once Magma reaches the surface, geologists identify it as lava.

COMPOSITION

Igneous rocks are classified according to how and where they formed, in other words, if they'replutonic or volcanic, and their mineral composition, describing the minerals they contain. Themineral compositions of igneous rocks are usually described as being felsic, intermediate, mafic, or ultramafic.

As a mafic magma starts to cool, some of the silica combines with iron and magnesium to make olivine. As it cools further, much of the remaining silica goes into calcium-rich plagioclase, and any silica left may be used to convert some of the olivine to pyroxene. Soon after that, all of the magma is used up and no further changes take place. The minerals present will be olivine, pyroxene, and calcium-rich plagioclase. If the magma cools slowly underground, the product will be gabbro; if it cools quickly at the surface, the product will be basalt.

Felsic magmas tend to be cooler than mafic magmas when crystallization begins (because they don't have to be as hot to remain liquid), and so they may start out crystallizing pyroxene (not olivine) and plagioclase. As cooling continues, the various reactions on the discontinuous branch will proceed because silica is abundant, the plagioclase will become increasingly sodium-rich, and eventually, potassium feldspar and quartz will form. Commonly even very felsic rocks will not have biotite or muscovite because they may not have enough aluminum or enough hydrogen to make the OH complexes that are necessary for mica minerals. Typical felsic rocks are granite and rhyolite.

The image below shows some common igneous rocks classified by mode of occurrence and mineral composition.

The rocks listed in the table above are the most common igneous rocks, but there are more than 700 different types of igneous rocks. Granite is perhaps the most useful one for humans. We use granite in many building materials and art. As discussed in the introduction to this lesson, pumice is commonly used for abrasives. Peridotite is sometimes mined for peridot, a type of gemstone used in jewelry. Diorite is extremely hard and is commonly used for art, as it was used extensively by ancient civilizations for vases and other decorative artwork.

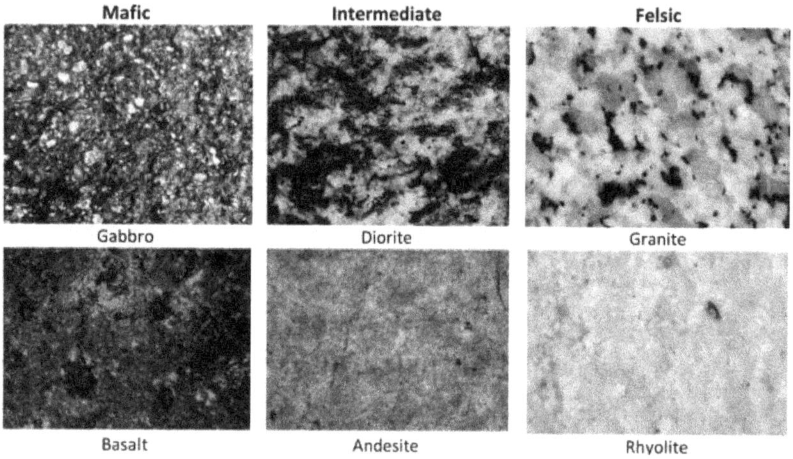

Figure 9.8 List of Common Igneous Rocks. Image is under a Creative Commons Attribution 4.0 International License.

SUMMARY

Igneous rocks form either when they cool very slowly deep within the Earth or when magma cools rapidly at the Earth's surface. The composition of the magma will determine the minerals that will crystallize forming different types of igneous rocks.

Intrusive, or plutonic, igneous rock forms when magma is trapped deep inside the Earth. Great globs of molten rock rise toward the surface. Some of the magma may feed volcanoes on the Earth's surface, but most remain trapped below, where it cools very slowly over many thousands or millions of years until it solidifies. Slow cooling means the individual mineral grains have a very long time to grow, so they grow to a relatively large size. Intrusive rocks have a coarse-grained texture.

Extrusive, or volcanic, igneous rock is produced when magma exits and cools above (or very near) the Earth's surface. These are the rocks that form at erupting volcanoes and oozing fissures. The magma, called lava when molten rock erupts on the surface, cools and solidifies almost instantly when it is exposed to the relatively cool temperature of the atmosphere. Quick cooling means that mineral crystals don't have much time to grow, so these rocks have a very fine-grained or even glassy texture. Hot gas bubbles are often trapped in the quenched lava, forming a bubbly, vesicular texture.

10 Chapter | SEDIMENTARY ROCKS

Figure 10.1 Red Rock State Park, in Cantil California. Image by Jeremy Patrich is used under a CC-BY 4.0 license.

Goals & Objectives of this unit

- Describe the differences between cobbles, pebbles, sand, silt, and clay and explain the relationship between clast size and the extent to which clasts can be transported by moving water or by the wind.
- Explain the differences in the characteristics and depositional environments of various types of chemical sedimentary rocks.
- Differentiate between various sedimentary depositional environments in both terrestrial and marine environments and explain how the formation of sedimentary basins can be related to plate tectonic processes.

Apply your understanding of the features of sedimentary rocks, including grain characteristics, sedimentary structures, and fossils, to the interpretation of past depositional environments and climates.

WHAT ARE SEDIMENTARY ROCKS?

Transportation is the movement of sediments or dissolved ions from the site of erosion to a site of deposition; this can be by wind, flowing water, glacial ice, or mass movement down a slope. The deposition takes place where the conditions change enough so that sediments being transported can no longer be transported, like when a current slows. Burial occurs when more sediments are piled onto existing sediments, and layers formed earlier are covered and compacted. Lithification is what happens at depths of hundreds to thousands of meters when those compacted sediments become cemented together to form solid sedimentary rock.

In this unit, we divide sedimentary rocks into two main types: clastic and chemical. Clastic sedimentary rocks are mainly composed of a material that has been transported as solid fragments (clasts). Chemical sedimentary rocks are mainly composed of a material that has been transported as ions in solution. It's important not to assume that mechanical weathering leads only to clastic sedimentary rocks, while chemical weathering leads only to chemical sedimentary rocks. In most cases, millions of years separate the weathering and depositional processes, and both types of sedimentary rocks tend to include at least some material derived from both types of weathering.

Clastic Sedimentary

A clast is a fragment of rock or mineral, ranging in size from less than a micron (too small to see) to as big as an apartment block. The smaller clasts tend to be composed of a single mineral crystal, and the larger ones are typically composed of pieces of rock. Most sand-sized clasts are made of quartz because quartz is more resistant to weathering than any other common mineral. Most clasts that are smaller than sand size (<1/16 mm) are made of clay minerals.

Most clasts larger than sand size (>2 mm) are actual fragments of rock, and commonly these might be fine-grained rock like basalt or andesite, or if they are bigger, coarse-grained rock like granite or gneiss.

GRAIN-SIZE CLASSIFICATION

There are six main grain-size categories; five are broken down into subcategories,0with clay being the exception. The diameter limits for each successive subcategory are twice as large as the one beneath it. In general, a boulder is bigger than a toaster and difficult to lift. There is no upper limit to the size of a boulder. A small cobble will fit in one hand, a large one in two hands. A pebble is something that you could throw quite easily. The smaller ones, known as granules, are gravel size, but still, you could throw one. But you can't throw a single grainof sand. Sand ranges from 2 mm down to 0.063 mm, and its key characteristic is that it feels sandy or gritty between your fingers even the finest sand grains feel that way. Silt is essentially too small for individual grains to be visible, and while sand feels sandy to your fingers, silt feelssmooth to your fingers but gritty in your mouth. Clay is so fine that it feels smooth even in your mouth.

By utilizing in-class models, students could identify features on in-class 3d models, which include; glacial, fluvial, and costal. Students could also assess how the features are formed and interpret materials and processes that shape the lithosphere, hydrosphere, atmosphere, and solar system.

Description		Size Range in mm		Size in microns	
		from	to	from	to
Boulder	large	1,024	no limit		
	medium	512	1024		
	small	256	512		
Cobble	large	128	256		
	small	64	128		
Pebble (Granule)	very coarse	32	64		
	coarse	16	32		
	medium	8	16		
	fine	4	8		
	very fine	2	4		
Sand	very coarse	1	2	1,000	2,000
	coarse	0.5	1	500	1,000
	medium	0.25	0.5	250	500
	fine	0.125	0.25	125	250
	very fine	0.063	0.125	63	125
Silt	very coarse			32	63
	coarse			16	32
	medium			8	16
	fine			4	8
	v. fine			2	4
Clay	clay			0	2

Figure 10.2 The Udden-Wentworth Grain-Size Scale for Classifying Sediments & the Grains that make up Sedimentary Rocks. Image is used under a Creative Commons Attribution 4.0 International License.

TRANSPORTATION

One of the key principles of sedimentary geology is that the ability of a moving medium (air or water) to move sedimentary particles and keep them moving, is dependent on the velocity of flow. The faster the medium flows, the larger the particles it can move. Parts of the river are moving faster than other parts, especially where the slope is greatest, and the channel is narrow. Not only does the velocity of a river change from place to place, but it changes from season to season.

If you drop a granule into a glass of water, it will sink quickly to the bottom (less than half a second). If you drop a grain of sand into the same glass, it will sink more slowly (a second or two depending on the size). A grain of silt will take several seconds to get to the bottom, and a particle of fine clay may never get there. The rate of settling is determined by the balance between gravity and friction.

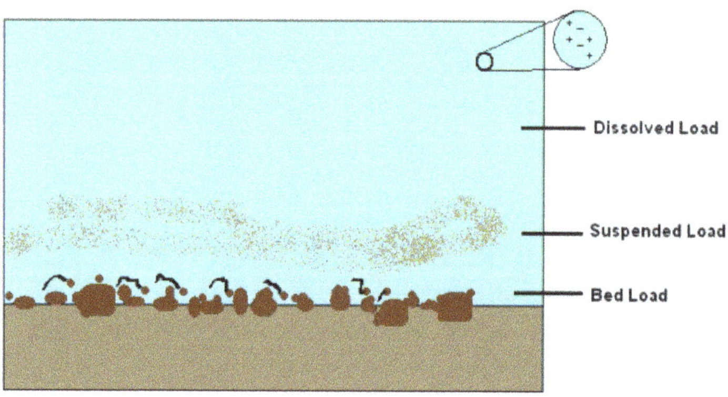

Figure 10.3 Examples of how Material Moves in A River Environment, 50% moves by Saltation, Bed/Traction & Dissolved/ Solution Load Equally Transport the Remaining 50%. Image is in the public domain.

During peak discharge at this location, the water is high enough to flow over the embankment on the right, and it flows fast enough to move the boulders that cannot be moved during low flows.

Large bedload clasts are pushed (by traction) or bounced along the bottom (saltation), while smaller clasts are suspended in the water and kept there by the turbulence of the flow. As the flow velocity changes, different-sized clasts may be either incorporated into the flow or deposited on the bottom. At various places along a river, there are always some clasts being deposited, some staying where they are, and some being eroded and transported. This changes over time as the discharge of the river changes in response to changing weather conditions.

Other sediment transportation media, such as waves, ocean currents, and wind, operate under similar principles, with flow velocity as the key underlying factor that controls transportation and deposition. Clastic sediments are deposited in a wide range of environments, including glaciers, slope failures, rivers, both fast and slow, lakes, deltas, and ocean environments, both shallow and deep. Depending on the grain size in particular, they may eventually form into rocks ranging from fine mudstone to coarse breccia and conglomerate.

Figure 10.4 An Example of Bedload-- Notice That the Clasts are too Large to Move. Image by Jeremy Patrich is used under a CC-BY 4.0 license.

LITHIFICATION

Lithification is the term used to describe several different processes that take place within a deposit of sediment to turn it into solid rock. One of these processes is burial by other sediments, which leads to compaction of the material and removal of some of the intervening water and air. After this stage, the individual clasts are all touching one another. Cementation is the process of crystallization of minerals within the pores between the small clasts, and also at the points of contact between the larger clasts (sand size and larger). Depending on the pressure, temperature, and chemical conditions, these crystals might include calcite, hematite, quartz, clay minerals, or a range of other minerals.

The characteristics and distinguishing features of clastic sedimentary rocks are summarized in the table 10.1.

Table 10.1 The Main Types of Clastic Sedimentary Rocks & Their Characteristics

GROUP	EXAMPLES	CHARACTERISTICS
MUDROCK	Mudstone Shale	> 75% silt and clay, not bedded > 75%
COAL		Dominated by fragments of partially decayed plant matter, often enclosed between beds of sandstone or mudrock

SANDSTONE	Quartz Sandstone Arkose Lithic Wacke	Dominated by sand, > 90% quartz Dominated by sand, > 10% feldspar Dominated by sand, > 10% rock fragments, > 15% silt and clay
CONGLOMERATE		Dominated by rounded clasts, pebble size and larger
BRECCIA		Dominated by angular clasts, pebble size and larger

Mudrock is composed of at least 75% silt- and clay-sized fragments. If it is dominated by clay, it is called claystone. If it shows evidence of bedding or fine laminations, it is shale; otherwise, it is mudstone. Mudrocks form in very low energy environments, such as lakes, river backwaters, and the deep ocean.

Most coal forms in fluvial or delta environments where vegetation growth is vigorous and where decaying plant matter accumulates in long-lasting swamps with low oxygen levels. To avoid oxidation and breakdown, the organic matter must remain submerged for centuries or millennia, until it is covered with another layer of either muddy or sandy sediments.

It is important to note that in some textbooks coal is described as an "organic sedimentary rock." In this book, coal is classified with the clastic rocks for two reasons: first, because it is made up of fragments of organic matter; and second, because coal seams (sedimentary layers) are almost always interbedded with layers of clastic rocks, such as mudrock or sandstone. In other words, coal accumulates in environments where other clastic rocks accumulate.

It's worth taking a closer look at the different types of sandstone because sandstone is a common and important sedimentary rock. The term arenite applies to a so-called clean sandstone, meaning one with less than 15% silt and clay. Considering the sand-sized grains only, arenites with 90% or more quartz are called quartz arenites. If they have more than 10% feldspar and more feldspar than rock fragments, they are called feldspathic arenites or arkosic arenites (or just arkose). If they have more than 10% rock fragments, and more rock fragments than feldspar, they are lithic arenites. A sandstone with more than 15% silt or clay is called a wacke, (pronounced wackie). The terms quartz wacke, lithic wacke, and feldspathic wacke are used. Another name for a lithic wacke is greywacke.

Clastic sedimentary rocks in which a significant proportion of the clasts are larger than 2 mm are known as conglomerate if the clasts are well rounded, and breccia if they are angular.

Conglomerates form in high-energy environments where the particles can become rounded, such as fast-flowing rivers. Breccia's typically formed where the particles are not transported a significant distance in the water, such as alluvial fans and talus slopes.

Figure 10.5 Breccia from Mosaic Canyon, Death Valley California. Clasts Size between 5cm- 1m. Image by Jeremy Patrich is used under a CC-BY 4.0 license.

CHEMICAL SEDIMENTARY ROCKS

Whereas clastic sedimentary rocks are dominated by components that have been transported as solid clasts (clay, silt, sand, etc.), chemical sedimentary rocks are dominated by components that have been transported as ions in solution (Na+, Ca2+, HCO3–, etc.). There is some overlap between the two because almost all clastic sedimentary rocks contain cement formed from dissolved ions, and many chemical sedimentary rocks include some clasts. Since ions can stay in solution for tens of thousands of years (some much longer) and can travel for tens of thousands of kilometers, it is virtually impossible to relate chemical sediments to their source rocks.

The most common chemical sedimentary rock, by far, is limestone. Others include chert, banded iron formations, and a variety of rocks that form when bodies of water evaporate. Biological processes are important in the formation of some chemical sedimentary rocks, especially limestone and chert. For example, limestone is made up almost entirely of fragments of marine organisms that manufacture calcite for their shells and other hard parts, and most chert includes at least some of the silica tests (shells) of tiny marine organisms such as diatoms.

Limestone

Almost all limestone forms in the oceans and most of that form on the shallow continental shelves, especially in tropical regions with coral reefs. Reefs are highly productive ecosystems populated by a wide range of organisms, many of which use calcium and bicarbonate ions in seawater to make carbonate minerals (especially calcite) for their shells and other structures. These include corals, of course, but also green and red algae, urchins, sponges, mollusks, and crustaceans. Especially after they die, but even while they are still alive, these organisms are eroded by waves and currents to produce carbonate fragments that accumulate in the surrounding region.

Reefs tend to form near the edges of steep drop-offs because the reef organisms thrive on nutrient-rich upwelling currents. As the reef builds up, it is eroded by waves and currents to produce carbonate sediments that are transported into the steep offshore fore-reef area and the shallower inshore back-reef area. These sediments are dominated by reef-type carbonate fragments of all sizes, including mud.

Calcite can also form on land in several environments. Tufa forms at springs and travertine (which is less porous) forms at hot springs. Similar material precipitates within limestone caves to form stalactites, stalagmites, and a wide range of other speleothems.

Figure 10.6 Tufa Towers in Mono Lake, California. Image by Jeremy Patrich is used under a CC-BY 4.0 license.

Dolomite

Dolomite is another carbonate mineral, but dolomite is also the name for a rock composed of the mineral dolomite (although some geologists use the term dolostone to avoid confusion). Dolomite rock is quite common (there's a whole

Italian mountain range named after it), which is surprising since marine organisms don't make dolomite. All of the dolomite found in ancient rocks has been formed through magnesium replacing some of the calcium in the calcite in carbonate muds and sands. This process is known as dolomitization, and it is thought to take place where magnesium-rich water percolates through the sediments in carbonate tidal flat environments.

Figure 10.7 Dolomite Formation in the Inyo Mountains, Notice the Native Pictograph. Image by Jeremy Patrich is used under a CC-BY 4.0 license.

Chert & Banded Iron Formations

As we've seen, not all marine organisms make their hard parts out of calcite; some, like diatoms, use silica, and when they die their tiny shells (or tests) settle slowly to the bottom where they accumulate as chert. In some cases, chert is deposited along with limestone in the moderately deep ocean, but the two tend to remain separate, so chert beds within limestone are quite common, as are nodules, link the flint nodules of the Cretaceous chalk of southeastern England. In other situations, and especially in very deep water, chert accumulates on its own, commonly in thin beds.

Some ancient chert beds, most dating to between 1800 and 2400 Ma, are also combined with a rock known as banded iron formation (BIF), a deep sea-floor deposit of iron oxide that is a common ore of iron. BIF forms when iron dissolved in seawater is oxidized, becomes insoluble, and sinks to the bottom in the same way that silica tests do to form chert. The prevalence of BIF in rocks dating from 2400 to 1800 Ma is due to the changes in the atmosphere and oceans that took place over that time period. Photosynthetic bacteria (e.g., cyanobacteria) consume carbon dioxide from the atmosphere and use solar energy to convert it to oxygen. These bacteria first evolved around 3500 Ma, and for the next billion years, almost all of that free oxygen was used up by chemical and biological processes, but by 2400 Ma free oxygen levels started to increase in the atmosphere and the oceans. Over a period of 600 million years, that oxygen gradually converted soluble ferrous iron to insoluble ferric iron, which combined with oxygen to form the mineral hematite, leading to the accumulation of BIFs. After 1800 Ma, little dissolved iron was left in the oceans and the formation of BIF essentially stopped.

Evaporites

In arid regions, lakes and inland seas typically have no stream outlet and the water that flows into them is removed only by evaporation. Under these conditions, the water becomes increasingly concentrated with dissolved salts, and eventually, some of these salts reach saturation levels and start to crystallize. Although all evaporate deposits are unique because of differences in the chemistry of the water, in most cases minor amounts of carbonates start to precipitate when the solution is reduced to about 50% of its original volume. Gypsum precipitates at about 20% of the original volume and halite precipitates at 10%. Other important evaporate minerals include borax.

Figure 10.8 Student Licking Halite from Badwater Basin, Death Valley California. Image by Jeremy Patrich is used under a CC-BY 4.0 license.

SEDIMENTARY STRUCTURES

Through careful observation over the past few centuries, geologists have discovered that the accumulation of sediments and sedimentary rocks takes place according to some important geological principles, as follows:

- The principle of original horizontality states that sediments accumulate in essentially horizontal layers. The implication is that tilted sedimentary layers observed today must have been subjected to tectonic forces.
- The principle of superposition states that sedimentary layers are deposited in sequence and that unless the entire sequence has been turned over by tectonic processes, the layers at the bottom are older than those at the top.
- The principle of inclusions states that any rock fragments in a sedimentary layer must beolder than the layer. For example, the cobbles in a conglomerate must have been formed before the conglomerate.
- The principle of faunal succession states that there is a well-defined order in which organisms have evolved through geological time, and therefore the identification of specific fossils in a rock can be used to determine its age.
- By understanding the origins of these features, we can make some very useful inferences about the processes that led to the deposition of the rocks that we are studying.

Bedding

Bedding, for example, is the separation of sediments into layers that either differs from one another in textures, composition, color, or weathering characteristics or are separatedby partings, narrow gaps between adjacent beds.

Figure 10.9 Example of Bedding, Afton Canyon, Mojave California. Image by Jeremy Patrich is used under a CC-BY4.0 license.

Bedding is an indication of changes in depositional processes that may be related to seasonal differences, changes in climate, changes in locations of rivers or deltas, or tectonic changes. Partings may represent periods of non-deposition that could range from a few decades to a few centuries. Bedding can form in almost any depositional environment.

Cross-Bedding

Cross-bedding is bedding that contains angled layers and forms when sediments are deposited by flowing water or wind. Cross-beds in streams tends to be on the scale of centimeters to tens of centimeters, while those in aeolian (wind deposited) sediments can be on the scale of meters to several meters.

Cross-bedded Jurassic Navajo Formation aeolian sandstone at Zion National Park, Utah. In most of the layers, the cross-beds dip down toward the right, implying wind direction from right to left during deposition. One bed dips in the opposite direction, implying an abnormal wind.

Cross-beds form as sediments is deposited on the leading edge of an advancing ripple or dune. Each layer is related to a different ripple that advances in the flow direction and is partially eroded by the following ripple. Cross-bedding is a very important sedimentary structure to

recognize because it can provide information on the direction of current flows and, when analyzed in detail, on other features like the rate of flow and the amount of sediment available.

Figure 10.10 Example of Cross-Bedding, Zion National Park Utah. Image by Steven Earle is under a CC By license.

Ripples

Ripples, which are associated with the formation of cross-bedding, may be preserved on the surfaces of sedimentary beds. Ripples can also help to determine flow direction as they tend to have their steepest surface facing downflow.

In a stream environment, boulders, cobbles, and pebbles can become imbricated, meaning that they are generally tilted in the same direction. Clasts in streams tend to tilt with their upper ends pointing downstream because this is the most stable position with respect to the streamflow.

Figure 10.11 Measuring Ripples, Then Compared Grain Size and Orientation to Lithified Ripples. Image by Jeremy Patrich is used under a CC-BY 4.0 license.

Mud Cracks

Mud cracks form when a shallow body of water (e.g., a tidal flat or pond), into which muddy sediments have been deposited, dries up and cracks. This happens because the clay in the upper mud layer tends to shrink on drying, and so it cracks because it occupies less space when it is dry.

Figure 10.12 Mud Cracks in Panamint Valley. Image by Jeremy Patrich is used under a CC-BY 4.0 license.

The various structures described above are critical to understanding and interpreting the formation of sedimentary rocks. In addition to these, geologists also look very closely at sedimentary grains to determine their mineralogy or lithology (in order to make inferences about the type of source rock and the weathering processes), their degree of rounding, their sizes, and the extent to which they have been sorted by transportation and depositional processes.

FOSSILS

The word fossil comes from the Latin term fossilis, meaning "dug up." Fossils are formed when an organism is buried by water containing debris and minerals and through the effects of wind or gravity. Most fossils are found in sedimentary rocks. Fossils can also be found in metamorphic rock or rock that has been altered by heat or pressure. Rarely are fossils found inigneous rock, which is formed when magma flows and hardens. The five most often cited types of fossils are mold, cast, imprint, permineralization and trace fossils.

Mold or Impression

A mold or impression fossil is formed when the plant or animal decays completely but leaves behind an impression of itself, like a hollow mold. No organic material is present and the organism itself is not copied. Mold or impression fossils can form in several ways, but generally enough air must be present to allow the organic material to completely decompose, which prevents the fossilization of the organism. These fossils are usually formed in sand or clay.

Cast Fossils

Cast fossils are the type people are most familiar with, as they make up the spectacular dinosaur skeletons on view in museums. Cast fossils occur when minerals deposit into the mold left by the rotting organic material, resulting in a three-dimensional replica of the hard structures of the plant or animal.

Figure 10.13 Imprint Fossil of an Ammonite, 65 Million Years Old. Image on Pixabay.

Imprint Fossils

Imprint fossils are found in silt or clay, like the mold or impression fossils, but they leave behind just a two-dimensional imprint. These fossils are sometimes found on exposed rock surfaces or when the layers in the rock are broken, revealing the fossil inside.

Permineralization

In permineralization, or petrified, fossils, each part of the organism is replaced by minerals, leaving a stone copy of the organism. Bones, teeth and even woody plant materials such as trees are sometimes preserved in this manner. One famous example of petrification is the hundreds of petrified trees in the Petrified Forest in Holbrook, Arizona.

Figure 10.14 A Fossilized Trilobite, Over 300 Million Years Old. Image by Jeremy Patrich is used under a CC-BY 4.0 license.

Trace Fossils

Trace fossils usually show tracks that animals made while moving across soft sediment. This sediment later hardens to become sedimentary rock. Trace fossils are valuable to paleontologists because by studying these footprints, scientists can discover how the animals moved, which in turn gives important information about the structure and even the life of the species.

SUMMARY

Sedimentary clasts are classified based on their size, and variations in clast size have important implications for transportation and deposition. Clastic sedimentary rocks range from conglomerate to mudstone. Clast size, sorting, composition, and shape are important features that allow us to differentiate clastic rocks and understand the processes that took place during their deposition.

Chemical sedimentary rocks form from ions that were transported in solution and then converted into minerals by biological and/or chemical processes. The most common chemical rock, limestone, typically forms in shallow tropical environments, where biological activity is a very important factor. Chert and banded iron formation are deep-ocean sedimentary rocks.

Evaporites form where the water of lakes and inland seas becomes supersaturated due to evaporation.

The deposition of sedimentary rocks takes place according to a series of important principles, including original horizontality, superposition, and faunal succession. Sedimentary rocks can also have distinctive structures that are important in determining their depositional environments. Fossils are useful for determining the age of a rock, the depositional environment, and the climate at the time of deposition.

11 Chapter

METAMORPHISM AND METAMORPHIC ROCKS

Figure 11.1 Polished Marbled in Mosaic Canyon, Death Valley California. Image by Jeremy Patrich is used under a CC-BY 4.0 license.

Goals & Objectives of this unit

- Summarize the factors that influence the nature of metamorphic rocks and explain why each one is important.
- Describe the mechanisms for the formation of foliation in metamorphic rocks.
- Classify metamorphic rocks based on their texture and mineral content and explain the origins of these differences.
- Describe the various settings in which metamorphic rocks are formed and explain thelinks between plate tectonics and metamorphism.

WHAT IS METAMORPHISM?

M is the change that takes place within a body of rock as a result of it being subjected to conditions that are different from those in which it formed. In most cases, but not all, this involves the rock being deeply buried beneath other rocks, where it is subjected to higher temperatures and pressures than those under which it formed. Metamorphic rocks typically have different mineral assemblages and different textures from their parent rocks, but they may have the same overall composition.

Most metamorphism results from the burial of igneous, sedimentary, or pre-existing metamorphic to the point where they experience different pressures and temperatures than those at which they formed. Metamorphism can also take place if cold rock near the surface is intruded and heated by a hot igneous body. Although most metamorphism involves temperatures above 150°C, some metamorphism takes place at temperatures lower than those at which the parent rock formed.

Controls over Metamorphic Processes

- The main factors that control metamorphic processes are:
- The mineral composition of the parent rock
- The temperature at which metamorphism takes place
- The amount and type of pressure during metamorphism
- The types of fluids (mostly water) that are present during metamorphism
- The amount of time available for metamorphism

PARENT ROCK

The parent rock is the rock that exists before metamorphism starts. In most cases, this is a sedimentary or igneous rock, but metamorphic rock that reaches the surface and is then reburied can also be considered a parent rock. On the other hand, if, for example, a mudstone is metamorphosed to slate and then buried deeper where it is metamorphosed to schist, the parent rock of the schist is mudstone, not slate. The critical feature of the parent rock is its mineral composition because it is the stability of minerals that counts when metamorphism takes place. In other words, when a rock is subjected to increased temperatures, certain minerals may become unstable and start to recrystallize into new minerals.

TEMPERATURE

The temperature that the rock is subjected to is a key variable in controlling the type of metamorphism that takes place. As we learned in the context of igneous rocks, mineral stability is a function of temperature, pressure, and the presence of fluids (especially water). All minerals are stable over a specific range of temperatures. For example, quartz is stable from environmental temperatures (whatever the weather can throw at it) all the way to 1800°C. If the pressure is higher, that upper limit will be higher. If there is water present, it will be lower.

On the other hand, most clay minerals are only stable up to about 150° or 200°C; above that, they transform into micas. Most other common minerals have upper limits between 150°C and 1000°C.

Some minerals will crystallize into different polymorphs (same composition, but different crystalline structure) depending on the temperature and pressure. Quartz is a good example as slightly different forms are stable between 0°C and 1800°C. The minerals kyanite, andalusite, and sillimanite are polymorphs with the composition Al_2SiO_5. They are stable at different pressures and temperatures, and, as we will see later, they are important indicators of pressures and temperatures in metamorphic rocks.

PRESSURE

Pressure is important in metamorphic processes for two main reasons. First, it has implications for mineral stability. Second, it has implications for the texture of metamorphic rocks. Rocks that are subjected to very high confining pressures are typically denser than others because the mineral grains are squeezed together, and because they may contain mineral polymorphs in which the atoms are more closely packed. Because of plate tectonics, pressures within the crust are typically not applied equally in all directions. In areas of plate convergence, the pressure in one direction (perpendicular to the direction of convergence) is typically greater than in the other directions. In situations where different blocks of the crust are being pushed in different directions, the rocks will be subjected to sheer stress. One of the results of directed pressure and sheer stress is that rocks become foliated, meaning that the ricks will have a directional fabric.

FLUIDS

Water is the main fluid present within rocks of the crust, and the only one that is considered here. The presence of water is important for two main reasons. First, water facilitates the transfer of ions between minerals and within minerals and therefore increases the rates at which metamorphic reactions take place. So, while the water doesn't necessarily change the outcome of a metamorphic process, it speeds the process up so metamorphism might take place over a shorter time period, or metamorphic processes that might not otherwise have had time to be completed are completed.

Secondly, water, especially hot water, can have elevated concentrations of dissolved substances, and therefore it is an important medium for moving certain elements around within the crust. So not only does water facilitate metamorphic reactions on a grain-to-grain basis, it also allows for the transportation of ions from one place to another. This is very important in hydrothermal processes, which are discussed toward the end of this chapter, and in the formation of mineral deposits.

TIME

Most metamorphic reactions take place at very slow rates. For example, the growth of new minerals within a rock during metamorphism has been estimated to be about 1 mm per million years. For this reason, it is very difficult to study metamorphic processes in a lab.

While the rate of metamorphism is slow, the tectonic processes that lead to metamorphism are also very slow, so in most cases, the chance for metamorphic reactions to be completed is high. For example, one important metamorphic setting is many kilometers deep within the roots of mountain ranges. A mountain range takes tens of millions of years to form, and tens of millions of years more to be eroded to the extent that we can see the rocks that were metamorphosed deep beneath it.

CLASSIFICATION OF METAMORPHIC ROCKS

There are two main types of metamorphic rocks: those that are foliated because they have formed in an environment with either directed pressure or shear stress, and those that are not foliated because they have formed in an environment without directed pressure or relatively near the surface with very little pressure at all. Some types of metamorphic rocks, such as quartzite and marble, which also form in directed-pressure situations, do not necessarily exhibit foliation because their minerals (quartz and calcite respectively) do not tend to show alignment.

When a rock is squeezed under directed pressure during metamorphism it is likely to be deformed, and this can result in a textural change such that the minerals are elongated in the direction perpendicular to the main stress. This contributes to the formation of foliation.

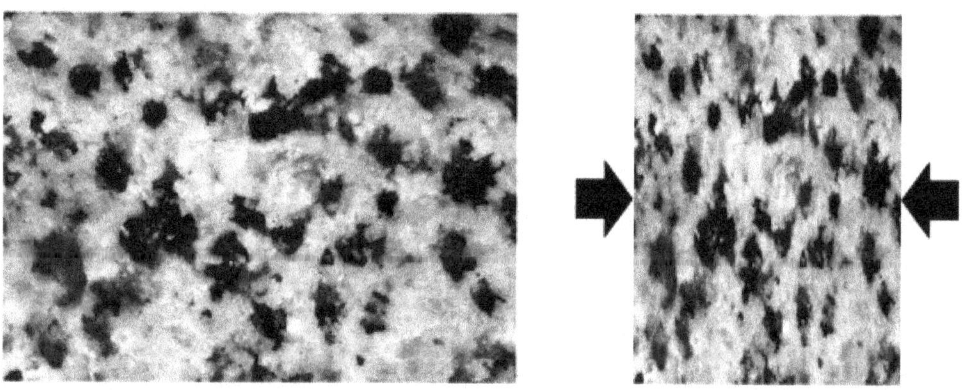

Figure 11.2 The Textural Effects of Squeezing During Metamorphism. Image by Steven Earle is used under a Creative Commons Attribution 4.0 International License.

When a rock is both heated and squeezed during metamorphism, and the temperature change is enough for new minerals to form from existing ones, there is a likelihood that the new minerals will be forced to grow with their long axes perpendicular to the direction of squeezing. After both heating and squeezing, new minerals have formed within the rock, generally parallel to each other, and the original bedding has been largely obliterated.

The various types of foliated metamorphic rocks, listed in order of the grade or intensity of metamorphism and the type of foliation are slate, phyllite, schist, and gneiss. As already noted, slate is formed from the low-grade metamorphism of shale, and has microscopic clay and mica crystals that have grown perpendicular to the stress. Slate tends to break into flat sheets.

Figure 11.3 Examples of Common Metamorphic Rocks. Image by Steven Earl and Michael C. Rygelis used under a CC-BY-SA.

Phyllite is similar to slate but has typically been heated to a higher temperature; the micas have grown larger and are visible as a sheen on the surface. Where slate is typically planar, phyllite can form in wavy layers. In the formation of schist, the temperature has been hot enough so that individual mica crystals are visible, and other mineral crystals, such as quartz, feldspar, or garnet may also be visible. In gneiss, the minerals may have separated into bands of different colors. In the example shown in Figure 7.8d, the dark bands are largely amphibole while the light-colored bands are feldspar and quartz. Most gneiss has little or no mica because it forms at temperatures higher than those under which micas are stable. Unlike slate and phyllite, which typically only form from mudrock, schist, and especially gneiss, can form from a variety of parent rocks, including mudrock, sandstone, conglomerate, and a range of both volcanic and intrusive igneous rocks.

Schist and gneiss can be identified based on both the unique and important minerals that are present. As an example, schist derived from basalt is typically rich in the mineral chlorite, so we call it chlorite schist. One derived from shale may be muscovite-biotite schist or just mica schist, or if there are garnets present it might be mica-garnet schist. Similarly, a gneiss that originated as basalt and is dominated by amphibole, is an amphibole gneiss or, more accurately, an amphibolite.

If a rock is buried to a great depth and encounters temperatures that are close to its melting point, it will partially melt. The resulting rock, which includes both metamorphosed and igneous material, is known as a migmatite.

As already noted, the nature of the parent rock controls the types of metamorphic rocks that can form from it under differing metamorphic conditions. The kinds of rocks that can be expected to form at different metamorphic grades from various parent rocks are listed in the table below.

Table 11.1 Metamorphic Rocks That Form from Different Parent Rocks

Approximate Temperature Ranges	Very Low Grade	Low Grade	Medium Grade	High Grade
PARENT ROCK	150 – 300°C	300 – 450°C	45 – 550°C	Above 550°C
MUDROCK	Slate	Phyllite	Schist	Gneiss
GRANITE	No change	No change	No change	Granite gneiss
BASALT	Chlorite schist	Chlorite schist	Amphibolite	Amphibolite
SANDSTONE	No change	Little change	Quartzite	Quartzite
LIMESTONE	Little change	Marble	Marble	Marble

Some rocks, such as granite, do not change much at the lower metamorphic grades because their minerals are still stable up to several hundred degrees.

Metamorphic rocks that form under either low-pressure conditions or just confining pressure do not become foliated. In most cases, this is because they are not buried deeply, and the heat for the metamorphism comes from a body of magma that has moved into the upper part of the crust. This is contact metamorphism. Some examples of non-foliated metamorphic rocks are marble, quartzite, and hornfels.

Marble is a metamorphosed limestone. When it forms, the calcite crystals tend to grow larger, and any sedimentary textures and fossils that might have been present are destroyed. If the original limestone was pure calcite, then the marble will likely be white, but if it had various impurities, such as clay, silica, or magnesium, the marble could be "marbled" in appearance.

Figure 10.4 An Example of Marble. Image is in the public domain.

Quartzite is metamorphosed sandstone. It is dominated by quartz, and in many cases, the original quartz grains of the sandstone are welded together with additional silica. Most sandstone contains some clay minerals and may also include other minerals such as feldspar or fragments of rock, so most quartzite has some impurities with the quartz.

Figure 10.5 AN Example of Quartzite. Image is in the public domain.

Hornfels is another non-foliated metamorphic rock that normally forms during contact metamorphism of fine-grained rocks like mudstone or volcanic rock. In some cases, hornfels has visible crystals of minerals like biotite or andalusite. If the hornfels formed in a situation without directed pressure, then these minerals would be randomly orientated, not foliated as they would be if formed with directed pressure.

Figure 11.6 An Example of Hornfels. Image is in the public domain.

PLATE TECTONICS & METAMORPHISM

All of the important processes of metamorphism that we are familiar with can be directly related to geological processes caused by plate tectonics.

Most regional metamorphism takes place within continental crust. While rocks can be metamorphosed at depth in most areas, the potential for metamorphism is greatest in the roots of mountain ranges where there is a strong likelihood for the burial of relatively young sedimentary rock to great depths. An example would be the Himalayan Range. At this continent-continent convergent boundary, sedimentary rocks have been both thrust up to great heights (nearly 9,000 m above sea level) and also buried to great depths. Considering that the normal geothermal gradient (the rate of increase in temperature with depth) is around 30°C per kilometer, rock buried to 9 km below sea level in this situation could be close to 18 km below the surface of the ground, and it is reasonable to expect temperatures up to 500°C. Metamorphic rocks formed there are likely to be foliated because of the strong directional pressure of converging plates.

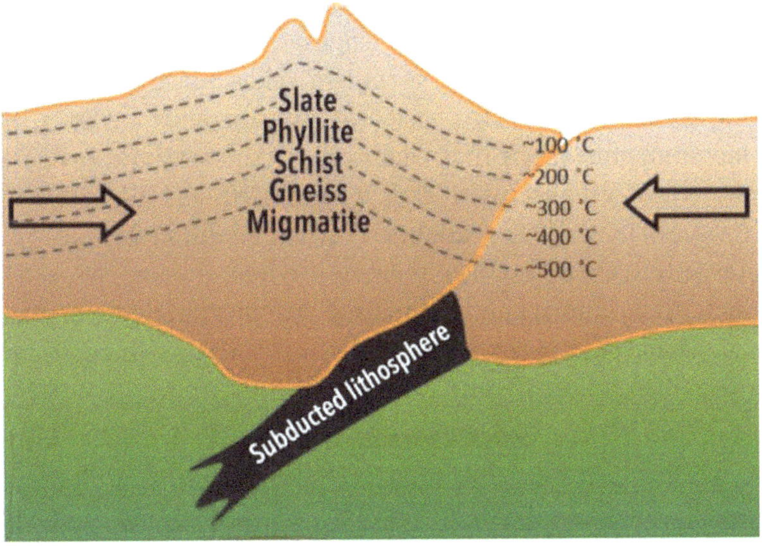

Figure 11.7 Regional Metamorphism beneath a Mountain Range. Image by Karla Panchuk (2018) CC BY 4.0, modified after Steven Earle (2015) CC BY 4.0.

At an oceanic spreading ridge, recently formed oceanic crust of gabbro and basalt is slowly moving away from the plate boundary. Water within the crust is forced to rise in the area close to the source of volcanic heat, and this draws more water in from farther out, which eventually creates a convective system where cold seawater is drawn into the crust and then out again onto the seafloor near the ridge.

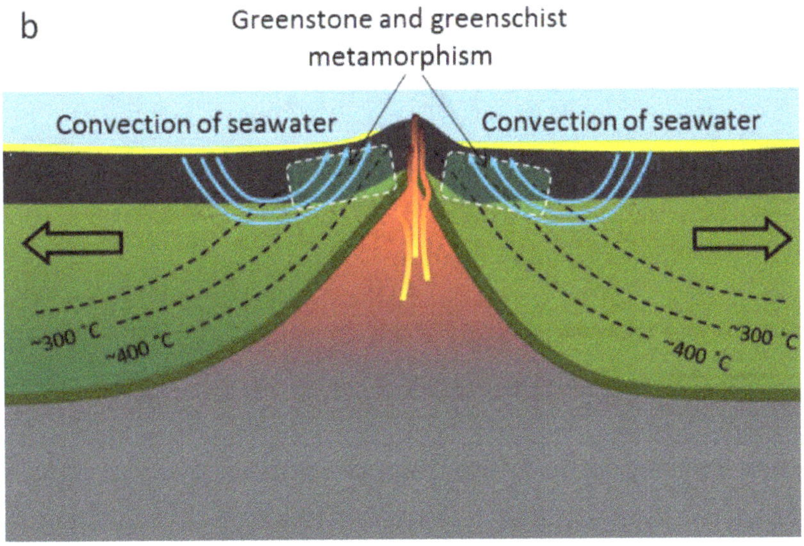

Figure 11.8 Regional Metamorphism of Oceanic Crustal Rock on Either Side of a Spreading Ridge. Image is used under a Creative Commons Attribution 4.0 International License.

The passage of this water through the oceanic crust at 200° to 300°C promotes metamorphic reactions that change the original pyroxene in the rock to chlorite and serpentine. Because this metamorphism takes place at temperatures well below the temperature at which the rock originally formed (~1200°C), it is known as retrograde metamorphism. The rock that forms in this way is known as greenstone if it isn't foliated, or greenschist if it is. Chlorite ((Mg5Al) (AlSi3)O10(OH)8) and serpentine ((Mg, Fe)3Si2O5(OH)4) are both "hydrated minerals" meaning that they have water (as OH) in their chemical formulas. When metamorphosed ocean crust is later subducted, the chlorite and serpentine are converted into new non-hydrous minerals (e.g., garnet and pyroxene) and the water that is released migrates into the overlying mantle, where it contributes to flux melting.

At a subduction zone, oceanic crust is forced down into the hot mantle. But because the oceanic crust is now relatively cool, especially along its sea-floor upper surface, it does not heat up quickly, and the subducting rock remains several hundreds of degrees cooler than the surrounding mantle. A special type of metamorphism takes place under these very high-pressure but relatively low-temperature conditions, producing an amphibole mineral is known as glaucophane (Na2(Mg3Al2)Si8O22(OH)2), which is blue and is a major component of a rock known as blueschist.

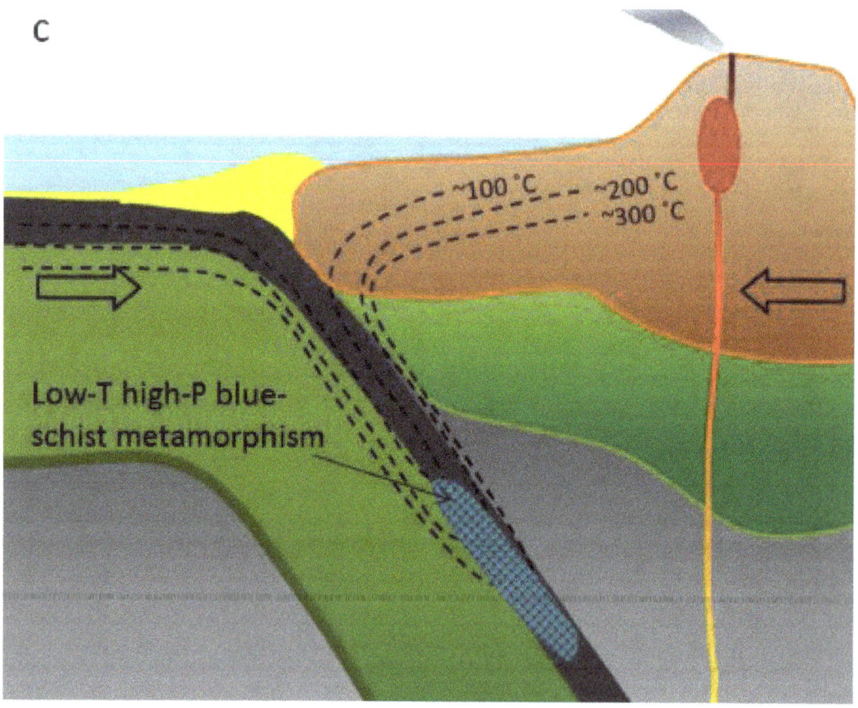

Figure 11.9 Regional Metamorphism of Oceanic Crust at a Subduction Zone. Image by Steven Earl is used under a Creative Commons Attribution 4.0 International License.

If you've never seen or even heard of blueschist, it's not surprising. What is surprising is that anyone has seen it! Most blueschist forms in subduction zones, continue to be subducted, turn into eclogite at about 35 km depth, and then eventually sinks deep into the mantle — never to be seen again. In only a few places in the world, where the subduction process has been interrupted by some tectonic process, has partially subductedblueschist rock returned to the surface. One such place is the area around San Francisco; the rock is known as the Franciscan Complex.

Magma is produced at convergent boundaries and rises toward the surface, where it can form magma bodies in the upper part of the crust. Such magma bodies, at temperatures of around 1000°C, heat up the surrounding rock, leading to contact metamorphism. Because this happens at relatively shallow depths, in the absence of directed pressure, the resulting rock does not normally develop foliation. The zone of contact metamorphism around an intrusion is very small (typically meters to tens of meters) compared with the extent of regional metamorphism in other settings (tens of thousands of square kilometers).

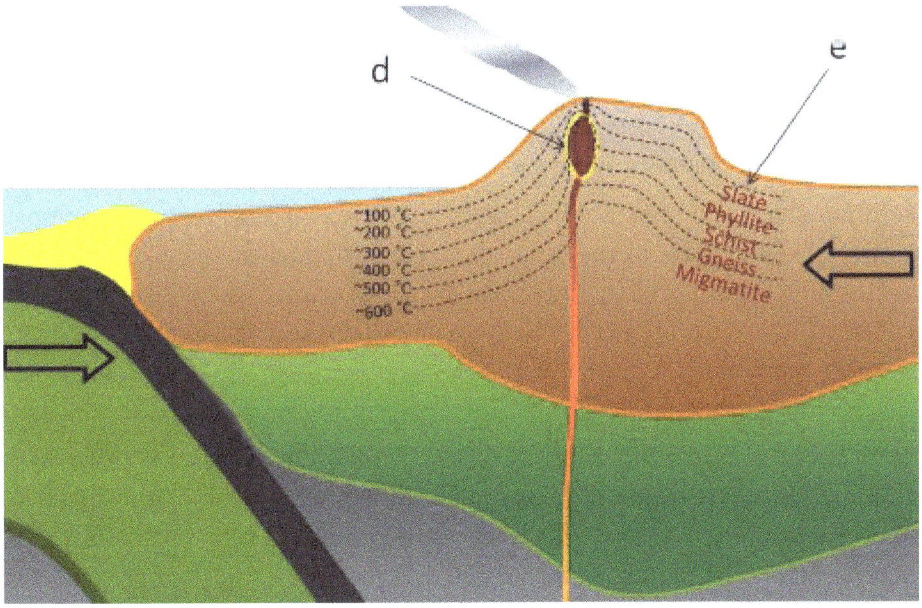

Figure 11.10 Contact Metamorphism around a High-Level Crustal Magma Chamber. Image by Steven Earl is used under a Creative Commons Attribution 4.0 International License.

Regional metamorphism also takes place within volcanic-arc mountain ranges, and because of the extra heat associated with the volcanism, the geothermal gradient is typically a little steeper in these settings (somewhere between 40° and 50°C/km). As a result, higher grades of metamorphism can take place closer to the surface than is the case in other areas.

Another way to understand metamorphism is by using a diagram that shows temperature on one axis and depth (which is equivalent to pressure) on the other. The three heavy dotted lines on this diagram represent Earth's geothermal gradients under different conditions. In most areas, the rate of increase in temperature with depth is 30°C/km. In other words, if you go 1,000 m down into a mine, the temperature will be roughly 30°C warmer than the average temperature at the surface. In most parts of southern Canada, the average surface temperature is about 10°C, so at 1,000 m depth, it will be about 40°C. That's uncomfortably hot, so deep mines must have effective ventilation systems. This typical geothermal gradient is shown by the green dotted line in Figure 7.20. At 10 km depth, the temperature is about 300°C and at 20 km it's about 600°C.

In volcanic areas, the geothermal gradient is more like 40° to 50°C/km, so the temperature at 10 km depth is in the 400° to 500°C range. Along subduction zones, as described above, the cold oceanic crust keeps temperatures low, so the gradient is typically less than 10°C/km.

SUMMARY

Metamorphism is controlled by five main factors: the composition of the parent rock, the temperature to which the rock is heated, the amount and type of pressure, the volumes and compositions of aqueous fluids that are present, and the amount of time available for metamorphic reactions to take place.

Metamorphic rocks are classified based on the texture and mineral composition. Foliation is a key feature of metamorphic rocks formed under directed pressure; foliated metamorphic rocks include slate, phyllite, schist, and gneiss. Metamorphic rocks formed in environments without strong directed pressure include hornfels, marble, and quartzite.

Almost all metamorphism can be explained by plate-tectonic processes. Oceanic crustal rock can be metamorphosed near the spreading ridge where it was formed, but most other regional metamorphism takes place in areas where mountain ranges have formed, which are most common at convergent boundaries. Contact metamorphism takes place around magma bodies in the upper part of the crust, which are also most common above convergent boundaries.

12 Chapter

WEATHERING AND SOILS

Figure 12.1 Highly Chemically Weathered Truck in Rhyolite, Nevada. Image by Jeremy Patrich is used under a CC-BY 4.0 license.

Goals & Objectives of this unit

- Explain why rocks formed at depth in the crust are susceptible to weathering at the surface.
- Describe the main processes of mechanical/physical weathering and the types of materials that are produced when mechanical weathering predominates.
- Describe the main processes of chemical weathering, and the products of chemical weathering of minerals such as feldspar, ferromagnesian silicates, and calcite.

- Discuss the relationships between weathering and soil formation, and the origins of soil horizons and some of the different types of soil.

WHAT IS WEATHERING?

Weathering is what takes place when a body of rock is exposed to the weather, in other words, to the forces and conditions that exist at Earth's surface. Except for volcanic rocks and some sedimentary rocks, most rocks are formed at some depth within the crust. There they experience relatively constant temperature, high pressure, no contact with the atmosphere, and little or no moving water. Once a rock is exposed at the surface, which is what happens when the overlying rock is eroded, conditions change dramatically. Temperatures vary widely, there is much less pressure, oxygen and other gases are plentiful, and in most climates, water is abundant.

Weathering includes two main processes that are quite different. One is the mechanical breakdown of rock into smaller fragments, and the other is the chemical change of the minerals within the rock to forms that are stable in the surface environment. Mechanical, also known as physical weathering, provides fresh surfaces for attack by chemical processes, and chemical weathering weakens the rock so that it is more susceptible to mechanical weathering.

Together, these processes create two very important products, one being the sedimentary clasts and ions in solution that can eventually become sedimentary rock, and the other being the soil that is necessary for our existence on Earth.

Mechanical Weathering

Intrusive igneous rocks form at depths of several hundreds of meters to several tens of kilometers. Sediments are turned into sedimentary rocks only when they are buried by other sediments to depths more than several hundreds of meters. Most metamorphic rocks are formed at depths of kilometers to tens of kilometers. Weathering cannot even begin until these rocks are uplifted through various processes of mountain building, most of which are related to plate tectonics, and the overlying material has been eroded and the rock is exposed as an outcrop.

- The important agents of mechanical weathering are:
- The decrease in pressure that results from removal of overlying rock.
- Freezing and thawing of water in cracks in the rock.
- Formation of salt crystals within the rock.
- Cracking from plant roots and exposure by burrowing animals.

When a mass of rock is exposed by weathering and removal of the overlying rock, there is a decrease in the confining pressure on the rock, and the rock expands. This unloading promotes cracking of the rock, known as exfoliation.

Granitic rock tends to exfoliate parallel to the exposed surface because the rock is typicallyhomogenous, and it doesn't have predetermined planes along which it must fracture. Sedimentary and metamorphic rocks, on the other hand, tend to exfoliate along predetermined planes.

FROST WEDGING

Frost wedging is the process by which water seeps into cracks in a rock, expands on freezing, and thus enlarges the cracks. The effectiveness of frost wedging is related to the frequency of freezing and thawing. Frost wedging is most effective in a climate like Canada's. In warm areas where freezing is infrequent, in very cold areas where thawing is infrequent, or in very dry areas, where there is little water to seep into cracks, the role of frost wedging is limited.

In many parts of the Sierra, the transition between freezing nighttime temperatures and thawing daytime temperatures is frequent, tens to hundreds of times a year. A common feature in areas of effective frost wedging is a talus slope, a fan-shaped deposit of fragments removed by frost wedging from the steep rocky slopes above

Figure 12.2 Example of Frost Wedging, Notice How the Rock is Cleanly Fractured. Image by Jeremy Patrich is used under a CC-BY 4.0 license.

EXFOLIATION

Exfoliation, separation of successive thin shells, or spalls, from massive rock such as granite or basalt; it is common in regions that have moderate rainfall. The thickness of theindividual sheet or plate may be from a few millimeters to a few meters.

Some geologists believe that exfoliation results when rocks formed at depth are exposed at the ground surface; the previous compressional forces would decrease and thus allow the rock to expand by fracturing parallel to the surface. Quite often, however, the fractures are not parallel to the ground surface, and this circumstance is taken as an indication of some other method of formation. Large daily variations in temperature, especially pronounced in deserts, were also credited with producing exfoliation, the expansion from heating during the day followed by contraction from rapid cooling at night was thought to cause the separation of thin slabs from large blocks of rock at the surface.

The study of thin shells that separate from rock exposed to the weather reveals a common cause of the separation of the slow development of clay minerals, which involves an increase in volume. The outer surface of exposed rock dries rapidly after wetting, but the moisture that penetrates minor crevices stays until some decay is started, and the resultant swelling causes flaking roughly parallel to the outer rock surface.

Figure 12.3 Exfoliated Granite Domes in the Alabama Hills, Lone Pine California. Image by Jeremy Patrich is used under a CC-BY 4.0 license.

Biologic Weathering

The effects of plants and animals are significant in biological weathering. Roots can force their way into even the tiniest cracks, and then they exert tremendous pressure on the rocks as they grow, widening the cracks and breaking the rock. Although animals do not normally burrow through solid rock, they can excavate and remove huge volumes of soil, and thus expose the rock to weathering by other mechanisms.

Figure 12.4 Example of Biological Weathering- Note How the Roots are Separating the Rock. Image by Jeremy Patrich is used under a CC-BY 4.0 license.

MASS WASTING

Mass wasting, also known as slope movement or mass movement, is the geomorphic process by which soil, sand, regolith, and rock move downslope typically as a solid, continuous or discontinuous mass, largely under the force of gravity, but frequently with characteristics of flow as in debris flows and mudflows. Types of mass wasting include creep, slides, flows, topples, and falls, each with its characteristic features, and taking place over timescales from seconds to hundreds of years. Mass wasting occurs on both terrestrial and submarine slopes and has been observed on Earth, Mars, Venus, and Jupiter's moon Io.

When the gravitational force acting on a slope exceeds its resisting force, slope failure (mass wasting) occurs. The slope material's strength and cohesion and the amount of internal friction between the materials help maintain the slope's stability and are known collectively as the slope's shear strength. The steepest angle that a cohesionless slope can maintain without losing its stability is known as its angle of repose. When a slope made of loose material possesses this angle, its shear strength perfectly counterbalances the force of gravity acting upon it.

Mass wasting may occur at a very slow rate, particularly in areas that are very dry or those areas that receive sufficient rainfall such that vegetation has stabilized the surface. It may also occur at very high speed, such as in rockslides or landslides, with disastrous consequences, both immediate and delayed.

Factors that change the potential of mass wasting include the change in slope angle, weakening of material by weathering, increased water content; changes in vegetation cover, and overloading. Sandwiched between a steep, unstable hillside (with the La Conchita Ranch Company situated on the plateau directly over the community), and the Pacific Ocean, La Conchita has been the site of recent major mudslides. In the image below- notice the different slides that have occurred. The arrows below the red line show additional slumps of material— perhaps an insight for future movements.

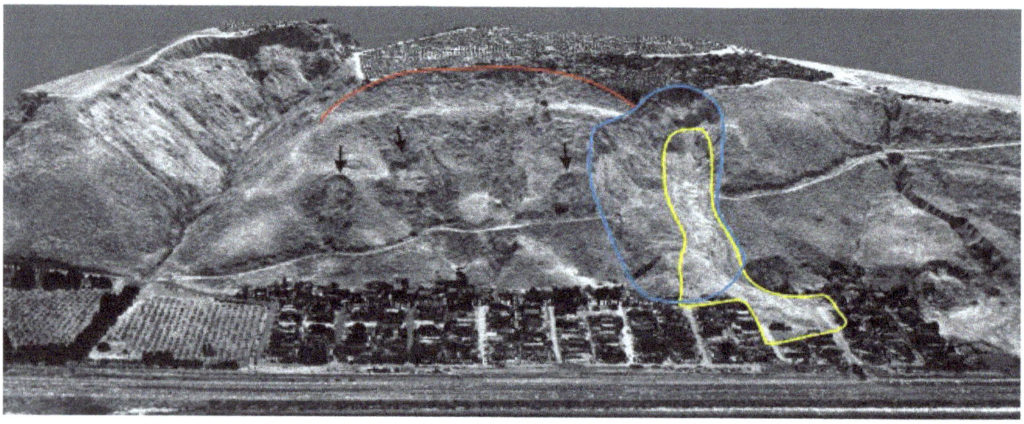

Figure 12.5 Landslides of La Conchita, Outlined in Blue is the 1995 Slide, Yellow is the 2005 Slide. Image courtesy of Airborne 1 Corporation, El Segundo, California is in the public domain.

Chemical Weathering

Chemical weathering results from chemical changes to minerals that become unstable when they are exposed to surface conditions. The kinds of changes that take place are highly specific to the mineral and the environmental conditions. Some minerals, like quartz, are virtually unaffected by chemical weathering, while others, like feldspar, are easily altered. In general, the degree of chemical weathering is greatest in warm and wet climates and least in cold and dry climates. The important characteristics of surface conditions that lead to chemical weathering are the presence of water (in the air and on the ground surface), the abundance of oxygen, and the presence of carbon dioxide, which produces weak carbonic acid when combined with water. That process, which is fundamental to most chemical weathering, can be shown as follows:

$$H2O + CO2 \longrightarrow H2CO3 \text{ then } H2CO3 \longrightarrow H^{+} + HCO3^{-},$$

water + carbon dioxide —> carbonic acid then carbonic acid —> hydrogen ion + carbonate ion

Here we have water, plus carbon dioxide in the atmosphere, combining to create carbonic acid. Then carbonic acid dissociates (comes apart) to form hydrogen and carbonate ions. The amount of CO2 in the air is enough to make only very weak carbonic acid, but there is typically much more CO2 in the soil, so water that percolates through the soil can become significantly more acidic.

There are two main types of chemical weathering. On the one hand, some minerals become altered to other minerals. For example, feldspar is altered, by hydrolysis, to clay minerals. On the other hand, some minerals dissolve completely, and their components go into solution. For example, calcite (CaCO3) is soluble in acidic solutions.

The hydrolysis of feldspar can be written like this:

CaAl2Si2O8 + H2CO3 + ½O2 —>Al2Si2O5(OH)4 + Ca^{2+} + $CO3^{2-}$

plagioclase + carbonic acid —> kaolinite + dissolved calcium + carbonate ions

This reaction shows calcium plagioclase feldspar, but similar reactions could also be written for sodium or potassium feldspars. In this case, we end up with the mineral kaolinite, along with calcium and carbonate ions in solution. Those ions can eventually combine (probably in theocean) to form the mineral calcite. Other silicate minerals can also go through hydrolysis, although the results will be a little different. For example, pyroxene can be converted to the clay minerals chlorite or smectite, and olivine can be converted to the clay mineral serpentine.

HYDROLYSIS

Whenever water reacts with another chemical compound, the process is called hydrolysis. Hydrolysis differs somewhat from hydration, although the two can occur together. Hydration is the bonding of whole water molecules to an ion (a charged atom or molecule), usuallya metal ion. Hydrolysis, on the other hand, involves an actual chemical reaction of the water molecule itself with another reactant. Aluminum ion, for example, can bond with six water molecules to form the hydrated aluminum ion. In water, however, the hydrated ion can undergo hydrolysis; some of the hydrated molecules contribute a hydrogen ion to the solution, making the solution acidic.

Solutions of non-hydrated ions often become either acidic or basic because of hydrolysis, too. In general, negative ions (anions) form basic solutions if they hydrolyze, because the negative charge on the ion attracts the positively charged hydrogen ion (H+) away from water, leaving the basic hydroxide ion (OH-) behind. Similarly, positive ions (cations) form acidic solutions if they hydrolyze, because the positive charge on the ion attracts the negatively charged hydroxide ion away from water, leaving the acidic hydrogen ion behind. Hydrolysis of these ions only occurs, however, if the ion originally came from a weak acid or base, or the salt of a weak acid or base. (A salt is an ionic chemical compound derived from an acid or base, often as the result of a neutralization reaction.) Ions do not hydrolyze if they are from strong acids or bases, such as chloride ion from hydrochloric acid or sodium ion from sodium hydroxide (a base) or their salts.

OXIDATION

Oxidation is another very important chemical weathering process. The oxidation of the iron in a ferromagnesian silicate starts with the dissolution of the iron. Picture below, a granitic rock containing biotite and amphibole which have been altered near to the rock's surface to limonite, which is a mixture of iron oxide minerals.

Figure 12.6 Example of Rusting or Oxidation on the links of a chain near the Golden Gate Bridge in San Francisco; it was continuously exposed to moisture and salt-laden spray, causing surface breakdown, cracking, and flaking of the metal. Image is used under a Attribution-Share Alike 3.0 Unported license.

For olivine, the process looks like this, where olivine in the presence of carbonic acid is converted to dissolved iron, carbonate, and silicic acid:

$$Fe_2SiO_4 + 4H_2CO_3 \longrightarrow 2Fe^{2+} + 4HCO_3^- + H_4SiO_4$$

Olivine + (carbonic acid) —> dissolved iron + dissolved carbonate + dissolved silicic acid In the presence of oxygen, the dissolved iron is then quickly converted to hematite:

$$2Fe^{2+} + 4HCO_3^- + \frac{1}{2} O_2 + 2H_2O \longrightarrow Fe_2O_3 + 4H_2CO_3$$

Dissolved iron + bicarbonate + oxygen + water—>hematite + carbonic acid

The equation shown here is for olivine, but it could apply to almost any other ferromagnesian silicate, including pyroxene, amphibole, or biotite. Iron in the sulfide minerals (e.g., pyrite) can also be oxidized in this way. The mineral hematite is not the only possible result, as there is a wide range of iron oxide minerals that can form in this way. The results of this process are illustrated below, which shows a granitic rock in which some of the biotite and amphibole has been altered to form the iron oxide mineral limonite.

ACID

Water contains many weak acids such as carbonic acid. This weak but abundant acid is formed when carbon dioxide gas from the atmosphere mixes with rainwater. Sulfur dioxide and nitrogen gases create other types of acid rain that act as chemical weathering agents. Some sources of sulfur dioxide are power plants that burn coal; as well as volcanoes and coastal marshes. Sulfur gases react with oxygen and rainwater to form sulfuric acid. Although relatively weak, acid's abundance and long-term effects produce noticeable damage to vegetation, fabrics, paints and, rocks. Below is a picture of the Parthenon, in the British Museum. In the 1930s, the curators at the museum decided there was something wrong with the surface of the marbles, and applied a caustic and abrasive chemical, leaving them as we see them today.

There may have been damaged from the polluted air of London at that time, but they removed any vestiges of the original painting that had been applied to the carvings.

Soil is a complex mixture of minerals (approximately 45%), organic matter (approximately 5%), and empty space (approximately 50%, filled to varying degrees with air and water). The mineral content of soils is variable but is dominated by clay minerals and quartz, along with minor amounts of feldspar and small fragments of rock. The types of weathering that take place within a region have a major influence on soil composition and texture. For example, in a warm climate, where chemical weathering dominates, soils tend to be richer in clay. Soil scientists describe soil texture in terms of the relative proportions of sand, silt, and clay, as shown in the soil triangle below. The sand and silt components in this diagram are dominated by quartz, with lesser amounts of feldspar and rock fragments, while the clay component is dominated by the clay minerals.

Figure 12.7 Chemical Weathering is observed on the Parthenon- 447-432BC. Image by Jeremy Patrich is used under a CC-BY 4.0 license.

FORMATION OF SOIL

Weathering is a key part of the process of the soil formation, and the soil is critical to our existence on Earth. In other words, we owe our existence to weathering, and we need to take care of our soil.

Many people refer to any loose material on Earth's surface as soil, but to earth scientists, soil is the material that includes organic matter, lies within the top few tens of centimeters of the surface, and is important in sustaining plant growth.

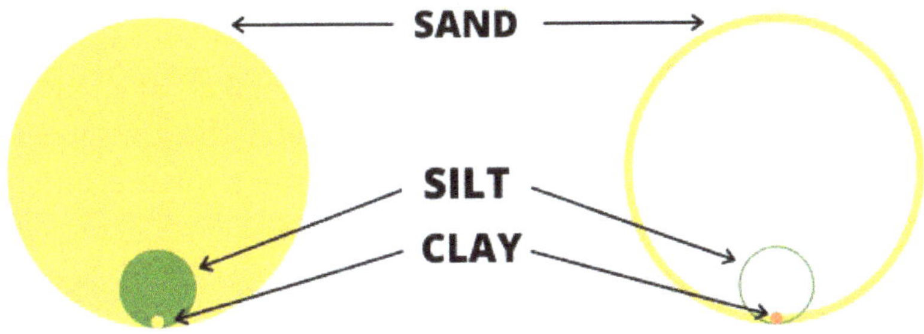

Figure 12.8 Size Comparison for Sand, Silt & Clay. Image by COC OER team is used under a CC BY 4.0 license.

The soil texture triangle gives names associated with various combinations of sand, silt, and clay. A coarse-textured or sandy soil is one comprised primarily of sand-sized particles. A fine- textured or clayey soil is one dominated by tiny clay particles. Due to the strong physical properties of clay, a soil with only 20% clay particles behaves as sticky, gummy clayey soil. The term loam refers to a soil with a combination of sand, silt, and clay-sized particles. For example, a soil with 30% clay, 50% sand, and 20% silt is called a sandy clay loam.

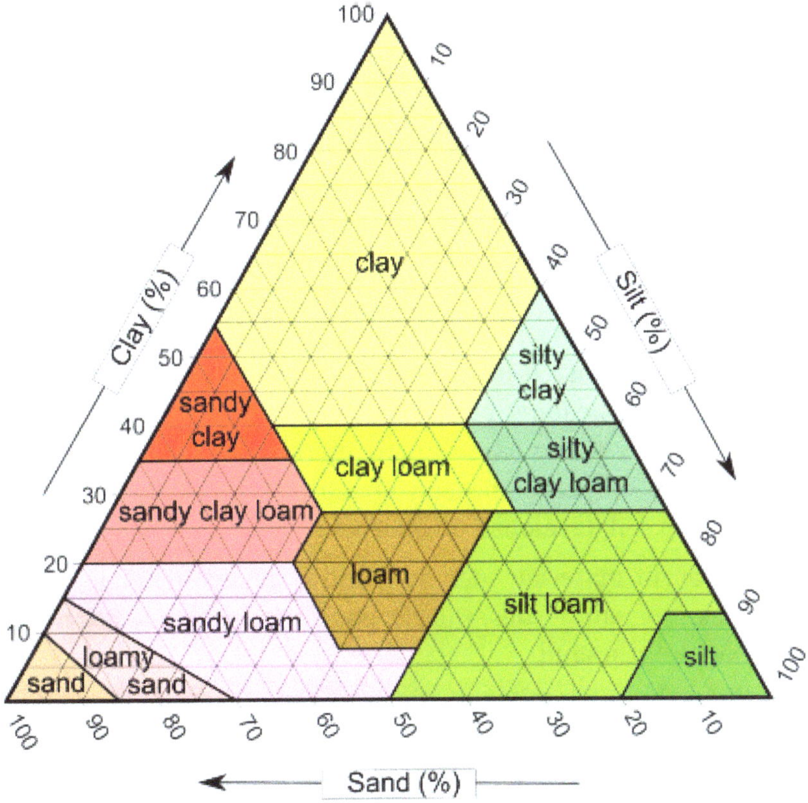

Figure 12.9 Soil Texture Triangle. Image by Mikenorton is used under a Attribution-Share Alike 3.0 Unported license.

Soil forms through the accumulation and decay of organic matter and both mechanical and chemical weathering processes described above. The factors that affect the nature of the soil and the rate of its formation include climate (especially average temperature and precipitation amounts, and the consequent types of vegetation), the type of parent material, the slope of the surface, and the amount of time available.

Climate

Soils develop because of the weathering of materials on Earth's surface, including the mechanical breakup of rocks, and the chemical weathering of minerals. Soil development is facilitated by the downward percolation of water. Soil forms most readily under temperate to tropical conditions (not cold) and where precipitation amounts are moderate (not dry, but not too wet). Chemical weathering reactions (especially the formation of clay minerals) and biochemical reactions proceed fastest under warm conditions, and plant growth is enhanced in warm climates. Too much water (e.g., in rainforests) can lead to the leaching of important chemical nutrients and hence to acidic soils. In humid and poorly drained regions, swampy conditions may prevail, producing soil that is dominated by organic matter. Too little water (e.g., in deserts and semi-deserts), results in very limited downward chemical transportationand the accumulation of salts and carbonate minerals (e.g., calcite) from upward-moving water. Soils in dry regions also suffer from a lack of organic material

Parent Material

Soil parent materials can include all different types of bedrock and any type of unconsolidated sediments, such as glacial deposits and stream deposits. Soils are described as residual soils if they develop on bedrock, and transported soils if they develop on transported material such as glacial sediments. But the term "transported soil" is misleading because it implies that the soil itself has been transported, which is not the case. When referring to such soil, it is better to be specific and say, "soil developed on unconsolidated material," because that distinguishes it from soil developed on bedrock.

Quartz-rich parent material, such as granite, sandstone, or loose sand, leads to the development of sandy soils. Quartz-poor material, such as shale or basalt, generates soils with little sand.

Parent materials provide important nutrients to residual soils. For example, a minor constituent of granitic rocks is the calcium-phosphate mineral apatite, which is a source of the important soil nutrient phosphorus. Basaltic parent material tends to generate very fertile soils because it also provides phosphorus, along with significant amounts of iron, magnesium, and calcium.

Some unconsolidated materials, such as river-flood deposits, make for especially good soils because they tend to be rich in clay minerals. Clay minerals have large surface areas with negative charges that are attractive to positively charged elements like calcium, magnesium, and iron, and potassium— important nutrients for plant growth.

Slope & Time

Soil can only develop where surface materials remain in place and are not frequently moved away by mass wasting. Soils cannot develop where the rate of soil formation is less than the rate of erosion, so steep slopes tend to have little or no soil.

Even under ideal conditions, the soil takes thousands of years to develop. As an example, most of northern California was still glaciated up until 10ka, and so, at that time, conditions were still not ideal for soil development even in the southern regions. Therefore, soils in northern California, are relatively young and not well developed.

The same applies to soils that are forming on newly created surfaces, such as recent deltas or sand bars, or in areas of mass wasting.

Soil Horizons

The process of soil formation generally involves the downward movement of clay, water, and dissolved ions, and a common result of that is the development of chemically and texturally different layers known as soil horizons. The typically developed soil horizons are:

- O: the layer of organic matter
- A: the layer of partially decayed organic matter mixed with mineral material
- E: the eluviated (leached) layer from which some of the clay and iron have beenremoved to create a pale layer that may be sandier than the other layers
- B: the layer of accumulation of clay, iron, and other elements from the overlying soil
- C: the layer of incomplete weathering
- R: the parent material or bedrock

Another type of layer that develops in hot arid regions (such as in the Mojave Desert) is known as caliche (pronounced ca-lee-chee). It forms from the downward (or in some cases upward) movement of calcium ions, and the precipitation of calcite within the soil. When well developed, caliche cements the surrounding material together to form a layer that has the consistency of concrete.

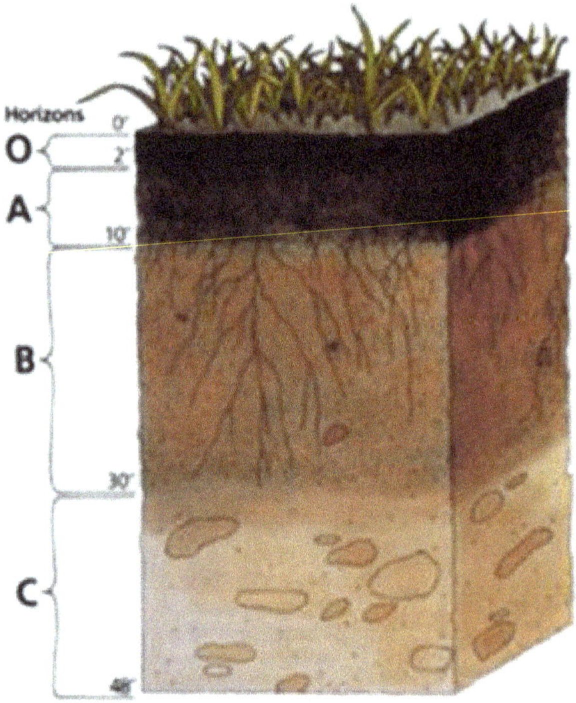

Figure 12.10 Soil Horizons. The R Horizon Would be Below the C Horizon. Image by USDA is in the public domain.

Like all geological materials, the soil is subject to erosion, although, under natural conditions on gentle slopes, the rate of soil formation either balances or exceeds the rate of erosion. Human practices related to forestry and agriculture have significantly upset this balance.

Soils are held in place by vegetation. When vegetation is removed, either through cutting trees or routinely harvesting crops and tilling the soil, that protection is either temporarily or permanently lost. The primary agents of the erosion of unprotected soil are water and wind.

Soil Erosion

Water erosion is accentuated on sloped surfaces because fast-flowing water has greater eroding power than still water. Raindrops can disaggregate exposed soil particles, putting the finer material (e.g., clays) into suspension in the water. Sheetwash, unchanneled flow across a surface carries suspended material away, and channels erode right through the soil layer, removing both fine and coarse material. Wind erosion is exacerbated by the removal of trees that act as windbreaks and by agricultural practices that leave bare soil exposed.

Tillage is also a factor in soil erosion, especially on slopes, because each time the soil is lifted by a cultivator, it is moved a few centimeters down the slope.

SUMMARY

Rocks weather when they are exposed to surface conditions, which in most cases are quite different from those at which they formed. The main processes of mechanical weathering include exfoliation, freeze-thaw, salt crystallization, and the effects of plant growth.

Chemical weathering takes place when minerals within rocks are not stable in their existing environment. Some of the important chemical weathering processes are hydrolysis of silicate minerals to form clay minerals, oxidation of iron in silicate and other minerals to form iron oxide minerals, and dissolution of calcite.

The main products of weathering and erosion are grains of quartz (because quartz is resistant to chemical weathering), clay minerals, iron oxide minerals, rock fragments, and a wide range of ions in solution.

Soil is a mixture of fine mineral fragments (including quartz and clay minerals), organic matter, and empty spaces that may be partially filled with water. Soil formation is controlled by climate (especially temperature and humidity), the nature and lithology of the parent material, the slope (because the soil can't accumulate on steep slopes), and the amount of time available. Typical soils have layers called horizons that form because of differences in the conditions with depth.

13 Chapter

EARTHS DYNAMIC SURFACE: PLATE TECTONICS

Figure 13.1 Viewing the Palmdale Road Cut, a 90 ft Slice Through Lakebed Sediments, Folded by The San Andreas Fault—The Meeting between the North American & Pacific Plates. Image by Jeremy Patrich is used under a CC-BY

Goals & Objectives of this unit

- Compare and describe each of these Earth layers: lithosphere, oceanic crust, and continental crust.
- Describe how convection takes place in the mantle and compare the two parts of thecore and describe why they are different from each other.
- Explain the concepts of the following hypothesis: continental drift hypothesis, seafloor spreading hypothesis, and the theory of plate tectonics.
- Describe the three types of tectonic plates, and how the processes lead to changes in
- Earth's surface features.

SEEING EARTH'S INTERIOR

Before you can learn about plate tectonics, you need to know something about the layers that are found inside Earth. These layers are divided by composition into core, mantle, and crust or by mechanical properties into the lithosphere and asthenosphere. Scientists use information from earthquakes and computer modeling to learn about Earth's interior. Humans have never drilled past Earth's crust, and yet we know a lot about the composition of the earth's interior. Rocks yield some clues, but they only reveal information about the outer crust. In rare instances, a mineral, such as a diamond, comes to the surface from deeper down in the crust or the mantle. To learn about Earth's interior, scientists use energy, recorded by seismographs, to see the different layers of the Earth, just like doctors can use an MRI, CT scan, or x-ray to see inside our bodies.

Seismic Waves

One ingenious way scientists learn about Earth's interior is by looking at how energy travels from the point of an earthquake, called seismic waves. Seismic waves travel outward in all directions from where the ground breaks at an earthquake. Seismograph stations measure the energy released by these earthquakes, but there are two that scientists are most interested regarding understanding the interior of the Earth.

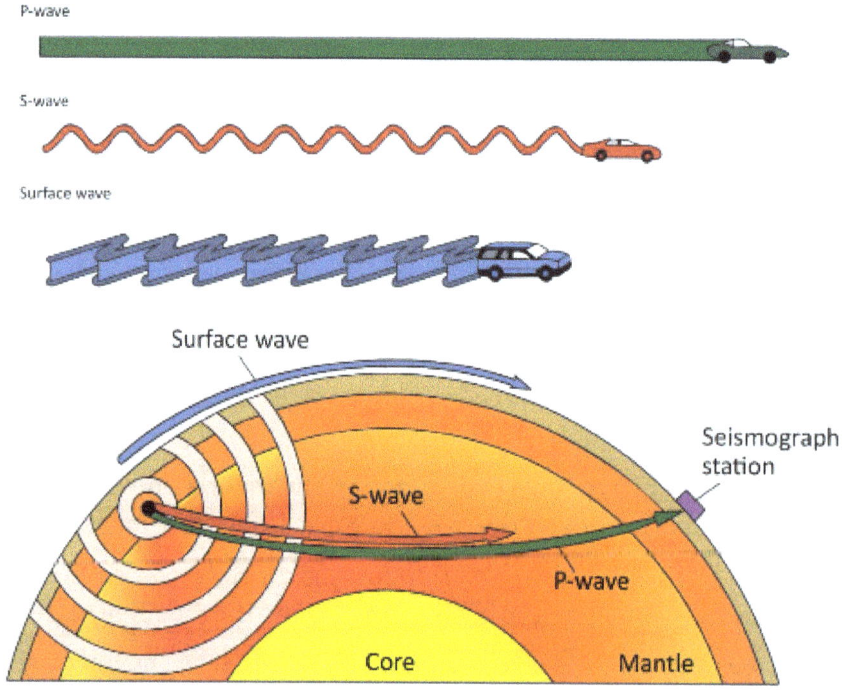

Figure 13.2 Example of P-Waves, S-Waves & Surface Waves. Image by COC OER team is used under a CC-BY4.0 license.

Primary waves (also called P-waves) are fastest, traveling at about 6 to 7 kilometers (about 4 miles) per second, so they arrive first at the seismometer. P-waves move in a compression or expansion type motion, squeezing and un-squeezing earth materials as they travel. P-waves bend slightly when they travel from one layer into another. Seismic waves move faster through denser or more rigid material. As P-waves encounter the liquid outer core, which is less rigid than the mantle, they slow down. This makes the P-waves arrive later and further away than would be expected. The result is a P-wave shadow zone. No P-waves are picked up at seismographs 104o to 140o from the earthquake's focus.

Secondary waves (also called S-waves) are about half as fast as P-waves, traveling at about 3.5 km (2 miles) per second, and arrive second at seismographs. S-waves move in an up and down motion perpendicular to the direction of wave travel. This produces a change in shape for the earth materials they move through. Only solids resist a change in shape, so S-waves are only able to propagate through solids. S-waves cannot travel through liquid.

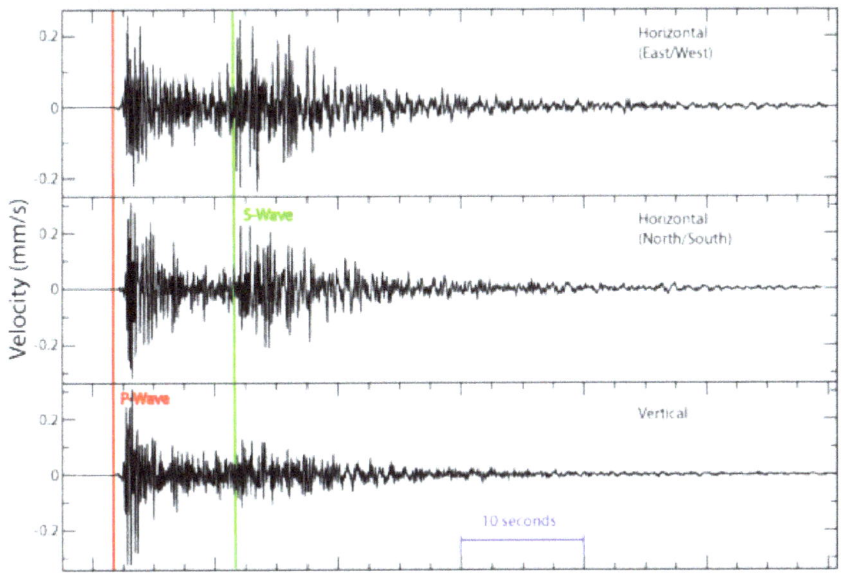

Figure 13.3 Velocity of Seismic Waves in the Earth, Versus Depth. Image under open copyright.

By tracking seismic waves, scientists have learned what makes up the planet's interior. P-waves slow down at the mantle core boundary, so we know the outer core is less rigid than the mantle. S-waves disappear at the mantle core boundary, so the outer core is liquid. Other clues to Earth's interior include the fact that we know that Earth's overall density is higher than the density of crustal rocks, so the core must be made of something dense, like metal. Also, since Earth has a magnetic field, there must be metal within the planet. Iron and nickel are both magnetic.

Finally, meteorites are the remains of the material that formed the early solar system and are thought to be similar to material in Earth's interior.

THE COMPOSITION & STRUCTURE OF EARTH

Core, mantle, and crust are divisions based on composition. The crust makes up less than 1%of Earth by mass, consisting of oceanic crust and continental crust is often more felsic rock. The mantle is hot and represents about 68% of Earth's mass. Finally, the core is mostly iron metal. The core makes up about 31% of the Earth. The lithosphere and asthenosphere are divisions based on mechanical properties. The lithosphere is composed of both the crust and the portion of the upper mantle that behaves as a brittle, rigid solid. The asthenosphere is partially molten upper mantle material that behaves plastically and can flow.

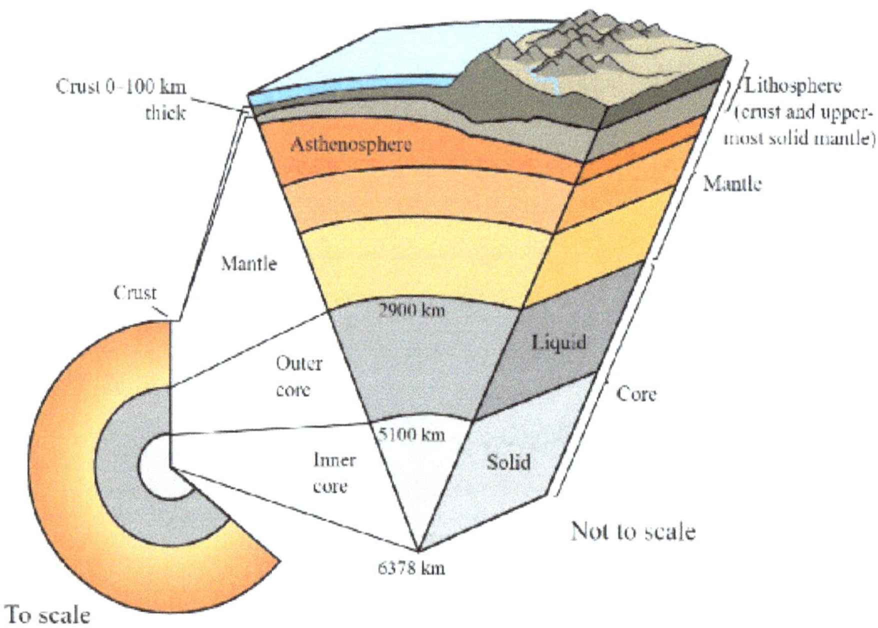

Figure 13.4 A Slice of Earth-- Exposing the Structure & Composition. Image is in the public domain.

Crust & Lithosphere

Earth's outer surface is its crust; a cold, thin, brittle outer shell made of rock. The crust is very thin, relative to the radius of the planet. There are two very different types of crust, each with its own distinctive physical and chemical properties.

Oceanic crust is composed of magma that erupts on the seafloor to create basalt lava flows or cools deeper down to create the intrusive igneous rock gabbro. Sediments, primarily muds and the shells of tiny sea creatures, coat the seafloor. Sediment is thickest near the shore where it comes off the continents in rivers and on wind currents. Continental crust is made up of many different types of igneous, metamorphic, and sedimentary rocks. The average composition is granite, which is much less dense than the mafic igneous rocks of the oceanic crust. Because it is thick and has relatively low density, continental crust rises higher on the mantle than oceanic crust, which sinks into the mantle to form basins. When filled with water, these basins form the planet's oceans. The lithosphere is the outermost mechanical layer, which behaves as a brittle, rigid solid. The lithosphere is about 100 kilometers thick. The definition of the lithosphere is based on how earth materials behave, so it includes the crust and the uppermost mantle, which are both brittle. Since it is rigid and brittle, when stresses act on the lithosphere, it breaks. This is what we experience as an earthquake.

Mantle

The two most important things about the mantle are:

- It is made of solid rock.
- It is hot.

Scientists know that the mantle is made of rock based on evidence from seismic waves, heat flow, and meteorites. The properties fit the ultramafic rock peridotite, which is made of the iron- and magnesium-rich silicate minerals. Peridotite is rarely found at Earth's surface.

Scientists know that the mantle is extremely hot because of the heat flowing outward from it and because of its physical properties. Heat flows in two different ways within the Earth: conduction and convection. Conduction is defined as the heat transfer that occurs through rapid collisions of atoms, which can only happen if the material is solid. Heat flows from warmer to cooler places until all are the same temperature. The mantle is hot mostly because of heat conducted from the core. Convection is the process of a material that can move, and flow may develop convection currents. Convection in the mantle is the same as convection in a pot of water on a stove. Convection currents within Earth's mantle form as material near the core heats up. As the core heats the bottom layer of mantle material, particles move more rapidly, decreasing its density and causing it to rise. The rising material begins the convection current. When the warm material reaches the surface, it spreads horizontally. The material cools because it is no longer near the core. It eventually becomes cool and dense enough to sink back down into the mantle. At the bottom of the mantle, the material travels horizontally and is heated by the core. It reaches the location where warm mantle material rises, and the mantle convection cell is complete.

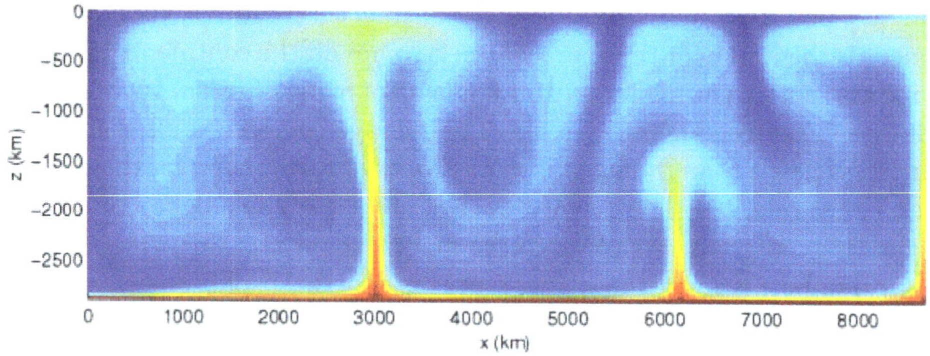

Figure 13.5 This is a Snapshot of One Time-Step in a Model of Mantel Convection. Image is used under a Attribution-Share Alike 3.0 Unported license.

Convection in the mantle is the same as convection in a pot of water on a stove. Convection currents within Earth's mantle form as material near the core heats up. As the core heats the bottom layer of mantle material, particles move more rapidly, decreasing its density and causing it to rise. The rising material begins the convection current. When the warm material reaches the surface, it spreads horizontally. The material cools because it is no longer near the core. It eventually becomes cool and dense enough to sink back down into the mantle. At the bottom of the mantle, the material travels horizontally and is heated by the core. It reaches the location where warm mantle material rises, and the mantle convection cell is complete.

Core

At the planet's center lies a dense metallic core. Scientists know that the core is metal for a few reasons. The density of Earth's surface layers is much less than the overall density of the planet, as calculated from the planet's rotation. If the surface layers are less dense than average, then the interior must be denser than average. Calculations indicate that the core is about 85% iron metal with nickel-metal making up much of the remaining 15%. Also, metallic meteorites are thought to be representative of the core. If Earth's core were not metal, the planet would not have a magnetic field. Metals such as iron are magnetic, but rock, which makes up the mantle and crust, is not. Scientists know that the outer core is liquid, and the inner core is solid because S waves stop at the inner core. The strong magnetic field is caused by convection in the liquid outer core. Convection currents in the outer core are due to heat from the even hotter inner core. The heat that keeps the outer core from solidifying is produced by the breakdown of radioactive elements in the inner core.

THEORY OF CONTINENTAL DRIFT

The continental drift hypothesis was developed in the early part of the 20th century, mostly by Alfred Wegener. Wegener said that continents move around on Earth's surface and that they were once joined together as a single supercontinent. While Wegener was alive, scientists did not believe that the continents could move. Find a map of the continents and cut each one out. Better yet, use a map where the edges of the continents show the continental shelf. That's the true size and shape of a continent and many can be pieced together like a puzzle. The easiest link is between the eastern Americas and western Africa and Europe, but the rest can fit together too.

Figure 13.6 Alfred Wegner a Few Years Before His Death in 1930. Image is in the public domain.

Alfred Wegener proposed that the continents were once united into a single supercontinent named Pangaea, meaning all earth in ancient Greek. He suggested that Pangaea broke up long ago and that the continents then moved to their current positions. He called his hypothesis continental drift.

Wegener pursued his theory with determination by combing the libraries, consulting with colleagues, and making observations, looking for evidence to support it. He relied heavily on matching geological patterns across oceans, such as sedimentary strata in South America matching those in Africa, North American coalfields matching those in Europe, and the mountains of Atlantic Canada matching those of northern Britain, both in morphology and rock type. Wegener also referred to the evidence for the Carboniferous and Permian (~300 Ma) Karoo Glaciation in South America, Africa, India, Antarctica, and Australia. He argued that this could only have happened if these continents were once all connected as a single supercontinent. He also cited evidence (based on his observations and interpretations) that showed that the continents were moving with respect to each other and determined a separation rate between Greenland and Scandinavia of 11 m per year, although he admitted that the measurements were not accurate. In fact, they weren't even close, the separation rate is about 2.5 cm per year.

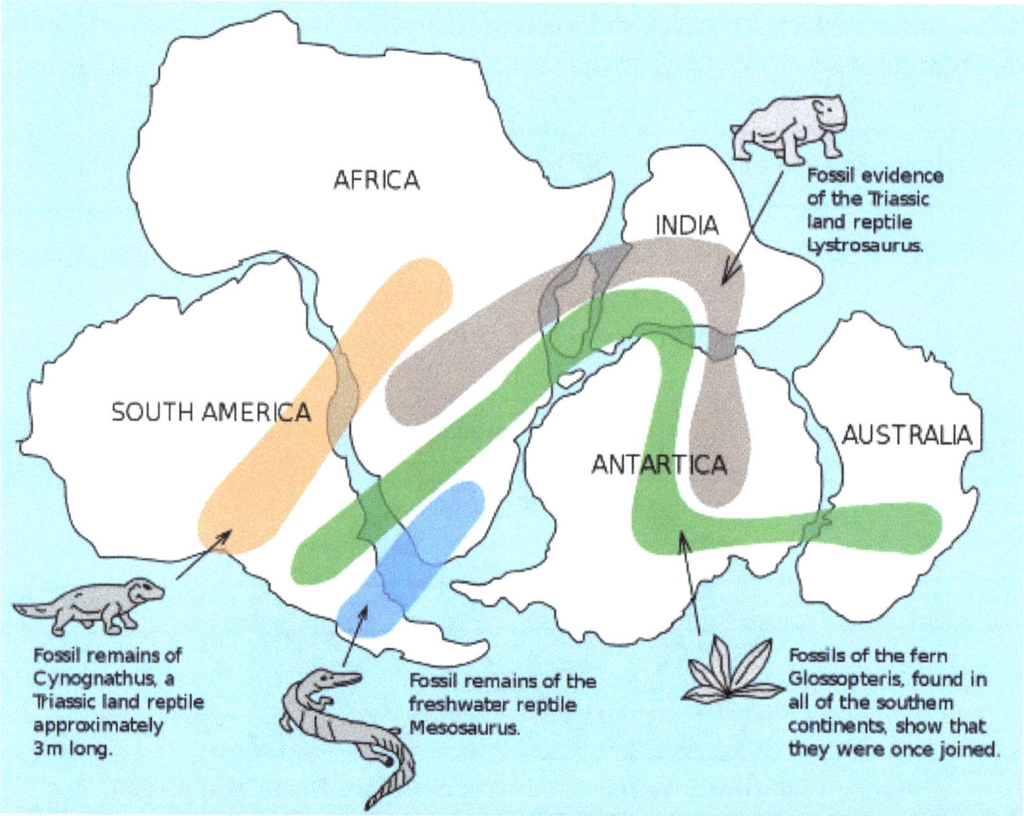

Figure 13.7 The Distribution of Several Permian Terrestrial Fossils that are Present in Various Parts of Continents that are now Separated by Oceans. Image by Steven Earl used under a CC-BY 4.0 international license.

Wegener first published his ideas in 1912 in a short book called Die Entstehung der Kontinente (The Origin of Continents), and then in 1915 in Die Entstehung der Kontinente undOzeane(The Origin of Continents and Oceans). He revised this book several times up to 1929. It was translated into French, English, Spanish, and Russian in 1924. Alfred Wegener died in Greenland in 1930 while carrying out studies related to glaciation and climate. At the time of his death, his ideas were tentatively accepted by only a small minority of geologists and soundly rejected by most. However, within a few decades, that was all to change.

Evidence for Continental Drift

Besides the way the continents fit together, Wegener and his supporters collected a great deal of evidence for the continental drift hypothesis. For one, identical rocks of the same type and age are found on both sides of the Atlantic Ocean. Wegener said the rocks had formed side-by- side and that the land had since moved apart. Mountain ranges with the same rock types, structures, and ages are now on opposite sides of the Atlantic Ocean. The Appalachians of the eastern United States and Canada, for example, are just like mountain ranges in eastern Greenland, Ireland, Great Britain, and Norway. Wegener concluded that they formed as a single mountain range that was separated as the continents drifted. Ancient fossils of the same species of extinct plants and animals are found in rocks of the same age but are on continents that are now widely separated. Wegener proposed that the organisms had lived side by side, but that the lands had moved apart after they were dead and fossilized. He suggested that the organisms would not have been able to travel across the oceans. For example, the fossils of the seed fern Glossopteris were too heavy to be carried so far by the wind. The reptile Mesosaurus could only swim in freshwater. Cynognathus and Lystrosaurus were land reptiles and were unable to swim. Rooves and rock deposits left by ancient glaciers are found today on different continents very close to the equator. This would indicate that the glaciers either formed in the middle of the ocean and/or covered most of the Earth. Today glaciers only form on land and nearer the poles. Wegener thought that the glaciers were centered over the southern landmass close to the South Pole and the continents moved to their present positions later on. Coral reefs and coal-forming swamps are found in tropical and subtropical environments, but ancient coal seams and coral reefs are found in locations where it is much too cold today. Wegener suggested that these creatures were alive in warm climate zones and that the fossils and coal later had drifted to new locations on the continents. Although Wegener's evidence was sound, most geologists at the time rejected his hypothesis of continental drift. Scientists argued that there was no way to explain how solid continents could plow through solid oceanic crust.

Wegener's idea was nearly forgotten until technological advances presented even more evidence that the continents moved and gave scientists the tools to develop a mechanism for Wegener's drifting continents.

Magnetic Polarity on The Same Continent with Rocks of Different Ages

Puzzling new evidence came in the 1950s from studies on the Earth's magnetic history. Scientists used magnetometers, devices capable of measuring the magnetic field intensity, to look at the magnetic properties of rocks in many locations. Geologists noted important things about the magnetic polarity of different aged rocks on the same continent. Magnetite crystals in fresh volcanic rocks point to the current magnetic north pole no matter what continent or where on the continent the rocks are located. Older rocks that are the same age and are located on the same continent point to the same location, but that location is not the current north magnetic pole. Older rocks that are of different ages do not point to the same locations or the current magnetic north pole. In other words, although the magnetite crystals were pointing to the magnetic north pole, the location of the pole seemed to wander. Scientists were amazed to find that the north magnetic pole changed location through time. There are three possible explanations for this:

- The continents remained fixed and the north magnetic pole moved.
- The north magnetic pole stood still, and the continents moved.
- Both the continents and the North Pole moved.

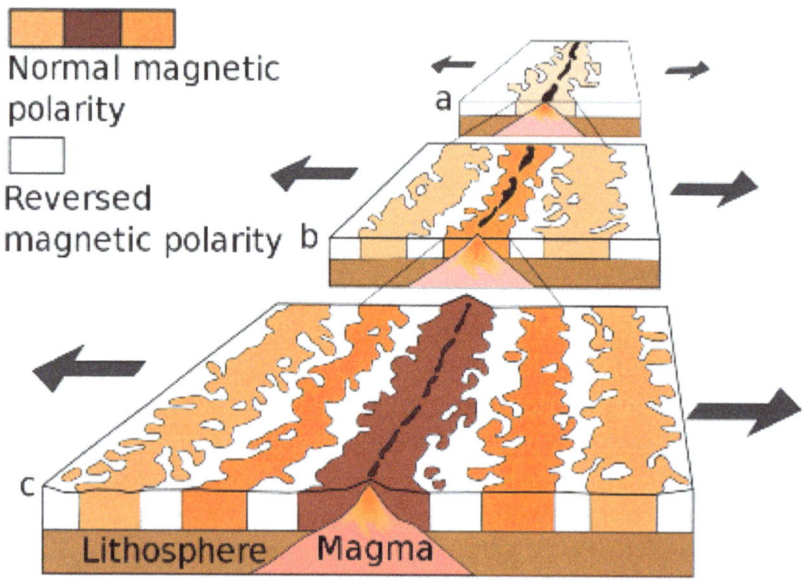

Figure 13.8 The Ocean Floor Shows Patterns of Magnetic Variation That is not Random, This is an Example of the Magnetic Stripping.
Image is in the public domain.

Magnetic Polarity on Different Continents with Rocks of The Same Age

Geologists noted that for rocks of the same age but on different continents, the little magnets pointed to different magnetic north poles. For example, 400-million-year-old magnetite in Europe pointed to a different north magnetic pole than the same-aged magnetite in North America. Around 250 million years ago, the north poles were also different for the two continents. The scientists looked again at the three possible explanations. Only one can be correct. If the continents had remained fixed while the north magnetic pole moved, there must have been two separate north poles. Since there is only one north pole today, the only reasonable explanation is that the north magnetic pole has remained fixed but that the continents have moved. To test this, geologists fitted the continents together as Wegener had done and behold, it worked. There has only been one magnetic north pole and the continents have drifted. They named the phenomenon of the magnetic pole that seemed to move but did not apparent polar wander. This evidence for continental drift gave geologists renewed interest in understanding how continents could move about on the planet's surface.

SEAFLOOR SPREADING

The discovery of magnetic striping naturally prompted more questions: How does the magnetic striping pattern form? And why are the stripes symmetrical around the crests of the mid-ocean ridges? These questions could not be answered without also knowing the significance of these ridges. In 1961, scientists began to theorize that mid-ocean ridges mark structurally weak zones where the ocean floor was being ripped in two lengthwise along the ridge crest. New magma from deep within the Earth rises easily through these weak zones and eventually erupts along the crest of the ridges to create a new oceanic crust. This process, later called seafloor spreading, operating over many millions of years has built the 50,000 km-long systems of mid- ocean ridges.

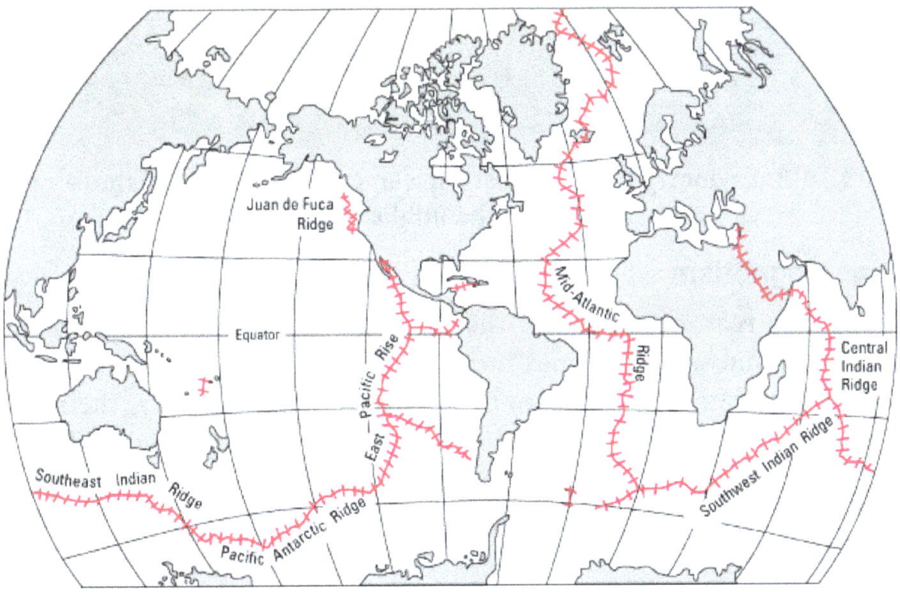

Figure 13.9 Simplified Map Showing Areas of Seafloor Spreading. Image is in the public domain.

Seafloor Bathymetry

World War II gave scientists the tools to find the mechanism for continental drift that had eluded Wegener. Maps and other data gathered during the war allowed scientists to develop the seafloor spreading hypothesis. This hypothesis traces oceanic crust from its origin at a mid- ocean ridge to its destruction at a deep-sea trench and is the mechanism for continental drift. During World War II, battleships and submarines carried echo sounders to locate enemy submarines. Echo sounders produce sound waves that travel outward in all directions, bounce off the nearest object, and then return to the ship. By knowing the speed of sound in seawater, scientists calculate the distance to the object based on the time it takes for the wave to make a round-trip. During the war, most of the sound waves ricocheted off the ocean bottom. After the war, scientists pieced together the ocean depths to produce bathymetric maps, which reveal the features of the ocean floor as if the water were taken away. Even scientists were amazed that the seafloor was not completely flat. What was discovered was a large chain of mountains along the deep seafloor, are called mid-ocean ridges. Scientists also discovered deep sea trenches along the edges of continents or in the sea near chains of active volcanoes. Finally, large, flat areas called abyssal plains we found. When they first observed these bathymetric maps, scientists wondered what had formed these features.

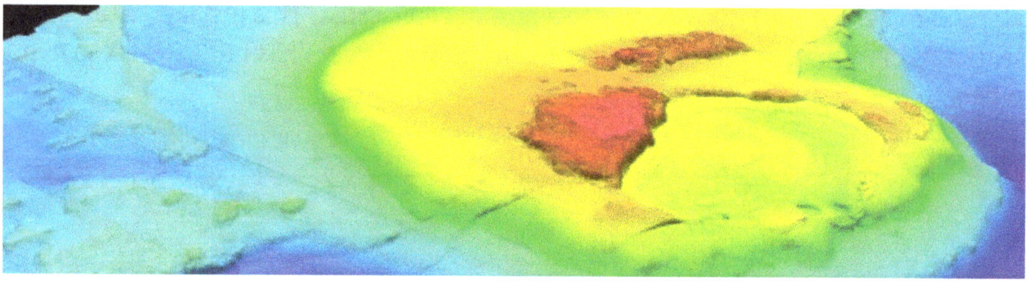

Figure 13.10 Bathymetry Map of East Flower Garden Bank, Offshore of Texas. Image is in the public domain.

Seafloor Magnetism

Sometimes, for reasons unknown, the magnetic poles switch positions. North becomes south and south becomes north. During normal polarity, the north and south poles are aligned as they are now. With reversed polarity, the north and south poles are in the opposite position. During WWII, magnetometers attached to ships to search for submarines located an astonishing feature; the normal and reversed magnetic polarity of seafloor basalts creates a pattern. Stripes of normal polarity and reversed polarity alternate across the ocean bottom. These stripes also form a mirror image of itself on either side of the mid-ocean ridges. But the stripes end abruptly at the edges of continents, sometimes at a deep-sea trench.

The oldest seafloor is near the edges of continents or deep-sea trenches and is less than 180 million years old. Since the oldest ocean crust is so much younger than the oldest continental crust, scientists realized that seafloor was being destroyed in a relatively short time.

Seafloor Spreading Hypothesis

Scientists brought these observations together in the early 1960s to create the seafloor spreading hypothesis. In this hypothesis, a hot buoyant mantle rises up a mid-ocean ridge, causing the ridge to rise upward. The hot magma at the ridge erupts as lava that forms new seafloor. When the lava cools, the magnetite crystals take on the current magnetic polarity and as more lava erupts, it pushes the seafloor horizontally away from the ridge axis. The magnetic stripes continue across the seafloor. As oceanic crust forms and spreads, moving away from the ridge crest, it pushes the continent away from the ridge axis. If the oceanic crust reaches a deep-sea trench, it sinks into the trench and is lost into the mantle. Scientists now know that the oldest crust is coldest and lies deepest in the ocean because it is less buoyant than the hot new crust. Seafloor spreading is the mechanism for Wegener's drifting continents. Convection currents within the mantle take the continents on a conveyor-belt ride of oceanic crust that over millions of years takes them around the planet's surface.

EARTH'S TECTONIC PLATES

When the concept of seafloor spreading came along, scientists recognized that it was the mechanism to explain how continents could move around Earth's surface. Scientific data and observation now allow us to merge the ideas of continental drift and seafloor spreading into the theory of plate tectonics. Seafloor and continents move around on Earth's surface, but what is moving? What portion of the Earth makes up the "plates" in plate tectonics? Thisquestion was also answered because of technology developed during the Cold War.

The plates are made up of the lithosphere. During the 1950s and early 1960s, scientists set up seismograph networks to see if enemy nations were testing atomic bombs.

These seismographs also recorded all of the earthquakes around the planet. The seismic records could be used to locate an earthquake's epicenter, the point on Earth's surface directly above the place where the earthquake occurs. Earthquake epicenters outline these tectonic plates. Mid-ocean ridges, trenches, and large faults mark the edges of these plates along with where earthquakes occur. The lithosphere is divided into a dozen major and several minor plates. The plates' edges can be drawn by connecting the dots that mark earthquakes' epicenters. A single plate can be made of all oceanic lithosphere or all continental lithosphere, but nearly all plates are made of a combination of both. Movement of the plates over Earth's surface is termed plate tectonics.

Plates move at a rate of a few centimeters a year, about the same rate fingernails grow.

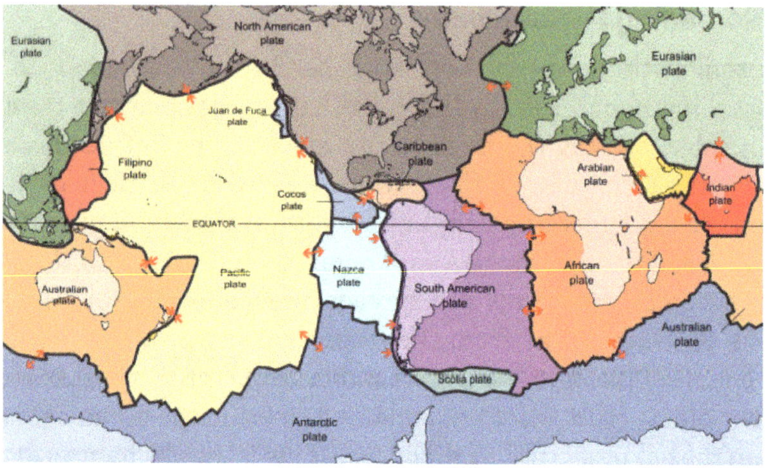

Figure 13.11 Simplified Map of Earth's Plates & Their Corresponding Tectonic Boundaries. Image is in the public domain.

How Plates Move

If seafloor spreading drives the plates, what drives seafloor spreading? Picture two convection cells side-by-side in the mantle. Hot mantle from the two adjacent cells rises at the ridge axis,creating new ocean crust. The top limb of the convection cell moves horizontally away from the ridge crest, as does the new seafloor.

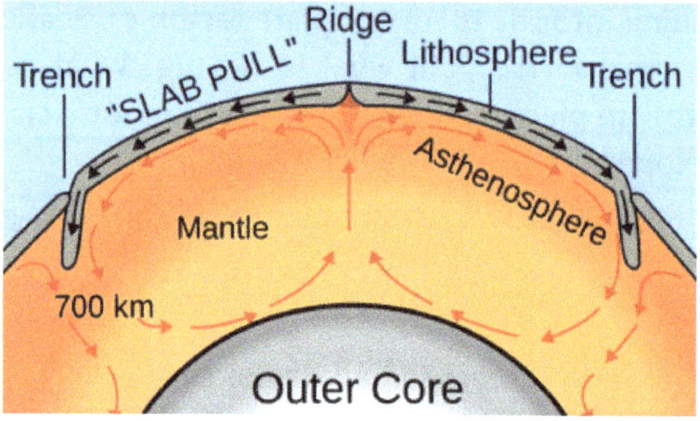

Figure 13.12 Example of Earth's Mantle Convection Cells. Image by Surachit is used under Attribution-Share Alike 3.0 Unported license.

The outer limbs of the convection cells plunge into the deeper mantle, dragging oceanic crust as well. This takes place at the deep-sea trenches. The material sinks to the core and moves horizontally. The material heats up and reaches the zone where it rises again.

TECTONIC BOUNDARIES

Plate boundaries are the edges where two plates meet. Most geologic activities, including volcanoes, earthquakes, and mountain building, take place at plate boundaries. How can two plates move relative to each other?

- Divergent plate boundaries: the two plates move away from each other.
- Convergent plate boundaries: the two plates move towards each other.
- Transform plate boundaries: the two plates slip past each other.

The type of plate boundary and the type of crust found on each side of the boundary determines what sort of geologic activity will be found there.

DIVERGENT

Plates move apart at mid-ocean ridges where new seafloor forms. Between the two plates is a rift valley. Lava flows at the surface cool rapidly to become basalt, but deeper in the crust, the magma cools more slowly to form gabbro. So the entire ridge system is made up of igneous rock that is either extrusive or intrusive. Earthquakes are common at mid-ocean ridges since the movement of magma and oceanic crust results in crustal shaking. The vast majority of mid- ocean ridges are located deep below the sea.

Figure 13.13 The Mid Atlantic Ridge. Image from Google Earth, Data SIO, NOAA, NGA, U.S. Navy, NGA, GEBCO: used under Google Earth reproduction guidelines.

As divergence occurs, shallow earthquakes can occur along with volcanoes along the rift areas. When the process begins, a valley will develop such as the Great Rift Valley in Africa. Over time that valley can fill up with water creating linear lakes. If divergence continues, a sea can form like the Red Sea and finally an ocean like the Atlantic Ocean. Check out the eastern half of Africa and notice the lakes that look linear. Eastern Africa is tearing apart from these linear lakes, to the Great Rift Valley, and up to the Red Sea. The ultimate divergent boundary is the Atlantic Ocean, which began when Pangea broke apart.

CONVERGENT

When two plates converge, the result depends on the type of lithosphere the plates are made of. No matter what, smashing two enormous slabs of lithosphere together results in the creation of magma and earthquakes. Ocean-to-continent convergence occurs when oceanic crust converges with continental crust, forcing the denser oceanic plate to plunge beneath thecontinental plate. This process called subduction occurs along oceanic trenchescalled subduction zones where lots of intense earthquakes and volcanic eruptions can occur. The denser, subducting plate begins to heat up under extreme pressure near the mantle and melts to create causes melting in the volcanoes. These coastal volcanic mountains are found in a line above the subducting plate. The volcanoes are known as a continental arc. The movement of crust and magma causes earthquakes.

USGS

The volcanoes of northeastern California, Lassen Peak, Mount Shasta, and Medicine Lake volcano are all along with the rest of the Cascade Mountains of the Pacific Northwest are the result of subduction of the Juan de Fuca plate beneath the North American plate. The Juan de Fuca plate is created by seafloor spreading just offshore at the Juan de Fuca Ridge. If the magma at a continental arc is felsic, it may be too viscous (thick) to rise through the crust. The magma will cool slowly to form granite or granodiorite. These large bodies of intrusive igneous rocks are called batholiths, which may someday be uplifted to form a mountain range.

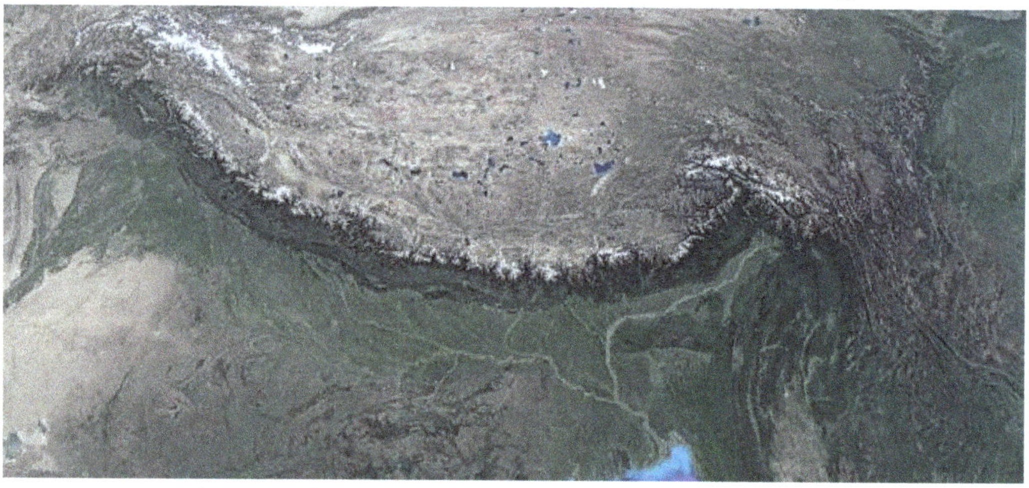

Figure 13.14 The Himalayan Mountains. Image from Google Earth, Data SIO, NOAA, NGA, U.S. Navy, NGA, GEBCO Landsat/Copernicus: used under Google Earth reproduction guidelines.

An oceanic-to-oceanic plate boundary occurs when two oceanic plates converge, causing the older, denser plate will subduct into the mantle. An ocean trench marks the location where the plate is pushed down into the mantle. The line of volcanoes that grows on the upper oceanic plate is an island arc. The Ring of Fire is a ring around the Pacific Ocean of subduction zones, which most are oceanic-to-oceanic convergence.

Along these subduction zones, volcanic islands (also called volcanic arcs) form. Examples of these regions include Japan, Indonesia, and the Aleutian Islands.

When two continental plates converge, instead of subduction, the two similar tectonic plates will buckle up to create large mountain ranges like a massive car pile-up. This iscalled continental-to-continental convergence, and geologically creates intense folding and faulting rather than volcanic activity.

Examples of mountain ranges created by this process are the Himalayan mountains as India is colliding with Asia, the Alps in Europe, and the Appalachian Mountains in the United States as the North American plate collided with the African plate when Pangea was forming. The Kashmir India earthquake of 2005 that killed over 80,000 people occurred because of this process. And most recently, the 2008 earthquake in China which killed nearly 85,000 people before the Summer Olympics was because of this tectonic force. The Appalachian Mountains are the remnants of a large mountain range that was created when North America rammed into Eurasia about 250 million years ago.

TRANSFORM

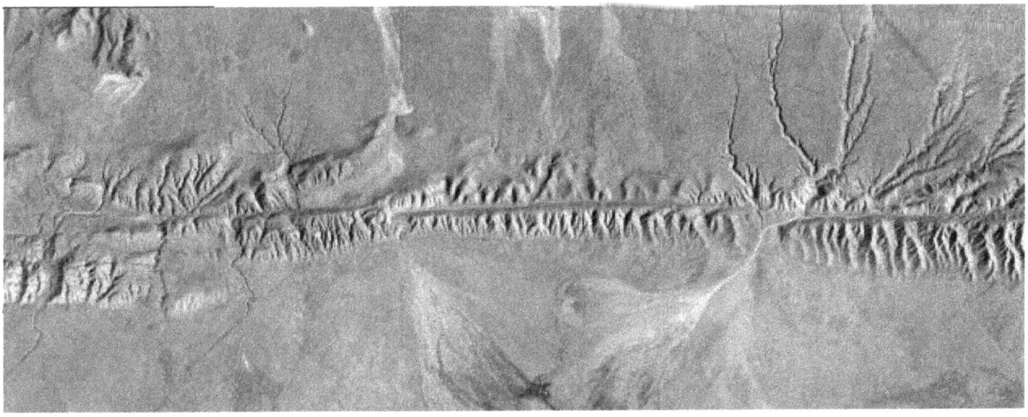

Figure 13.15 The San Andreas Fault (GoogleEarth). Image from Google Earth: used under Google Earth reproduction guidelines.

INTERPLATE BOUNDARIES

A small amount of geologic activity, known as an intraplate activity, does not take place at plate boundaries but within a plate instead. Mantle plumes are pipes of hot rock that rise through the mantle. The release of pressure causes melting near the surface to form a hotspot.

Eruptions at the hotspot create a volcano. Hotspot volcanoes are found in a line. Can you figure out why? Hint: The youngest volcano sits above the hotspot and volcanoes become older with distance from the hotspot. Geologists use some hotspot chains to tell the direction and the speed a plate is moving. Hotspot magmas rarely penetrate through thick continental crust. One exception is the Yellowstone hotspot.

SUMMARY

The evidence for continental drift in the early 20th century included the matching of continental shapes on either side of the Atlantic and the geological and fossil matchups between continents that are now thousands of kilometers apart.

The established theories of global geology were permanentism and contractionism, but neither of these theories was able to explain some of the evidence that supported the idea of continental drift.

Earth's lithosphere is made up of over 20 plates that are moving in different directions at rates of between 1 cm/y and 10 cm/y. The three types of plate boundaries are divergent (plates moving apart and new crust forming), convergent (plates moving together, and one being subducted), and transform (plates moving side by side). Divergent boundaries form where existing plates are rifted apart, and it is hypothesized that this is caused by a series of mantle plumes. Subduction zones are assumed to form where the accumulation of sediment at a passive margin leads to the separation of oceanic and continental lithosphere. Supercontinents form and break up through these processes.

It is widely believed that ridge-push and slab-pull are the main mechanisms for plate motion, as opposed to traction by mantle convection. Mantle convection is a key factor for producing the conditions necessary for ridge-push and slab-pull.

14 Chapter EARTHS DYNAMIC SURFACE: TECTONICS FORCE

Figure 14.1 Tilting of Material in Golden Canyon, Death Valley California. Image by Jeremy Patrich is used under a CC-BY 4.0 license.

Goals & Objectives of this unit

- Describe the causes of tectonic stress and how they influence earthquakes and volcanoes.
- Understand where earthquakes and volcanoes are located around the world and why.

- Explain geologic structures that lead to folding and faulting of the physical landscape.
- Determine how the composition of magma determines the explosiveness of the volcanic eruption.
- Describe the various types of fault lines and volcanoes and the landforms they form.

CAUSES & TYPES OF TECTONIC STRESS

Enormous slabs of lithosphere move unevenly over the planet's spherical surface, resulting in earthquakes. This chapter deals with two types of geological activity that occur because of plate tectonics: mountain building and earthquakes. First, we will consider what can happen to rocks when they are exposed to stress. Stress is the force applied to an object. In geosciences, stress is the force per unit area that is placed on a rock. Four types of stresses acting on materials.

- A deeply buried rock is pushed down by the weight of all the material above it. Since the rock cannot move, it cannot deform called confining stress.
- Compression squeezes rocks together, causing rocks to fold or fracture. Compression is the most common stress at convergent plate boundaries.
- Rocks that are pulled apart are under tension. Rocks under tension lengthen or break apart. Tension is the major type of stress at divergent plate boundaries.
- When forces are parallel but moving in opposite directions, the stress is called shear. Shear stress is the most common stress at transform plate boundaries.

When stress causes a material to change shape, it has undergone strain or deformation. Deformed rocks are common in geologically active areas. A rock's response to stress depends on the rock type, the surrounding temperature, and pressure conditions the rock is under the length of time the rock is under stress, and the type of stress. The rocks then have three possible responses to increasing stress: elastic deformation, plastic deformation, or fracturing.

Elastic deformation occurs when the rock returns to its original shape when the stress is removed. When rocks under stress do not return to its original shape when the stress is removed, it is called plastic deformation. Finally, when a rock under stress breaks, it's called a fracture.

Under what conditions do you think a rock is more likely to fracture? Is it more likely to break deep within Earth's crust or at the surface? What if the stress applied is sharp rather than gradual? At the Earth's surface, rocks usually break quite quickly, but deeper in the crust, where temperatures and pressures are higher, rocks are more likely to deform plastically. Sudden stress, such as a hit with a hammer, is more likely to make a rock break. Stress applied over time often leads to plastic deformation.

GEOLOGIC STRUCTURES

Sedimentary rocks are important for deciphering the geologic history of a region because they follow certain rules. First, sedimentary rocks are formed with the oldest layers on the bottom and the youngest on top. Second, sediments are deposited horizontally, so sedimentary rock layers are originally horizontal, as are some volcanic rocks, such as ash falls. Finally, sedimentary rock layers that are not horizontal are deformed in some manner. Often looking like they are tiling into the earth.

You can trace the deformation a rock has experienced by seeing how it differs from its original horizontal, oldest-on-bottom position. This deformation produces geologic structures such as folds, joints, and faults that are caused by stresses.

Folds

Rocks deforming plastically under compressive stresses crumple into folds. They do not return to their original shape. If the rocks experience more stress, they may undergo more folding or even fracture. There are three major types of rock folding: monoclines, synclines, and anticlines. A monocline is a simple bend in the rock layers so that they are no longer horizontal.

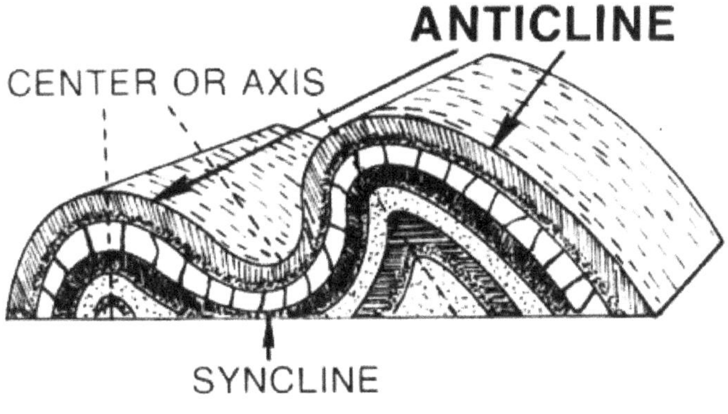

Figure 14.2 Example of Both a Syncline & Anticline. Image is used under a CC BY-SA: Attribution-ShareAlike license.

Faults

A rock under enough stress will eventually fracture. If there is no movement on either side of a fracture, the fracture is called a joint. But if the blocks of rock on one or both sides of a fracture move, the fracture is called a fault. Sudden motions along faults cause rocks to break and move suddenly, releasing the stored-up stress energy to create an earthquake. A slip is the distance rocks move along a fault and can be up or down the fault plane. Slip is relative because there is usually no way to know whether both sides moved or only one. Faults lie at an angle to the horizontal surface of the Earth. That angle is called the fault's dip. The dip defines which of two basic types a fault is. If the fault's dip is inclined relative to the horizontal, the fault is a dip-slip fault. There are two types of dip-slip faults. In normal faults, the hanging wall drops down relative to the footwall. Normal faults can be huge and are often responsible for uplifting mountain ranges in regions experiencing tensional stress. With reverse faults, the footwall drops down relative to the hanging wall. A type of reverse fault is a thrust fault, in which the fault plane angle is nearly horizontal. Rocks can slip many miles along thrust faults. A strike-slip fault is a dip-slip fault in which the dip of the fault plane is vertical and result from shear stresses. California's San Andreas Fault is the world's most famous strike-slip fault. It is a right- lateral strike-slip fault.

Figure 14.3 The Offset-- or Faulting in Red Rock State Park, in Cantil California. Image by Jeremy Patrich is used under a CC-BY 4.0 license.

Stress & Mountain Building

It is the shear power and strength of two or more converging continental plates smash upwards that create mountain ranges. Stresses from this uplift cause folds, reverse faults, and thrust faults, which allow the crust to rise upwards. The subduction of oceanic lithosphere at convergent plate boundaries also builds mountain ranges. When tensional stresses pull crust apart, it breaks into blocks that slide up and drop down along normal faults. The result is alternating mountains and valleys, known as a basin-and-range.

Causes of Earthquakes

An earthquake is a sudden ground movement caused by the sudden release of energy stored in rocks, called the elastic rebound theory. Earthquakes happen when so much stress builds up in the rocks that the rocks rupture. The energy is transmitted by seismic waves. Each year there are more than 150,000 earthquakes strong enough to be felt by people and 900,000 recorded by seismometers.

In an earthquake, the initial point where the rocks rupture in the crust is called the focus (sometimes called the hypocenter). The epicenter is the point on the land surface that is directly above the focus. In about 75% of earthquakes, the focus is in the top 10 to 15 kilometers (6 to 9 miles) of the crust. Shallow earthquakes cause the most damage because the focus is near where people live. However, it is the epicenter of an earthquake that is reported by scientists and the media.

EARTHQUAKES & ZONES

Nearly 95% of all earthquakes take place along one of the three types of tectonic plate boundaries, but earthquakes do occur along all three types of plate boundaries. About 80% of all earthquakes strike around the Pacific Ocean basin because it is lined with convergent and transform boundaries. Called the Ring of Fire, this is also the location of most volcanoes around the planet. About 15% take place in the Mediterranean-Asiatic Belt, where convergence is causing the Indian Plate to run into the Eurasian Plate creating the largest mountain ranges in the world. The remaining 5% are scattered around other plate boundaries or are intraplate earthquakes.

Transform Plate Boundary Zone

Transform plate boundaries occur where two tectonic plates are grinding parallel to each other rather than colliding or subducting. Deadly earthquakes occur at transform plate boundaries, creating strike-slip faults because they tend to have shallow focuses where the rupture occurs.

The faults along the San Andreas Fault zone produce around 10,000 earthquakes a year. Most are tiny, but occasionally one is massive. In the San Francisco Bay

Area, the Hayward Fault was the site of a magnitude 7.0 earthquake in 1868. The 1906 quake on the San Andreas Fault had a magnitude estimated at 7.9. During the 1989 World Series, a magnitude 7.1 earthquake struck Loma Prieta, near Santa Cruz, California, killing 63 people, injuring 3,756, and cost $6 billion. A few years later in Northridge, California, a magnitude 6.7 earthquake killed 72 people, injured 12,000 people, and caused $12.5 billion in damage. This earthquake occurred on an unknown fault because it was a blind thrust fault near Los Angeles, California. Although California is prone to many natural hazards, including volcanic eruptions at Mt. Shasta or Mt. Lassen, and landslides on coastal cliffs, the natural hazard the state is linked with is earthquakes. New Zealand also has strike-slip earthquakes, about 20,000 a year, but only a small percentage of those are large enough to be felt.

Figure 14.4 Strike-Slip Mode. Image is in the public domain.

Convergent Plate Boundary Zone

Earthquakes at convergent plate boundaries mark the motions of the subducting lithosphere as it plunges through the mantle, creating reverse and thrust faults. Convergent plate boundaries produce earthquakes all around the Pacific Ocean basin. The Philippine Plate and the Pacific Plate subduct beneath Japan, creating a chain of volcanoes and produces as many as 1,500 earthquakes annually. In March 2011 an enormous 9.0 earthquake struck off of Sendai in northeastern Japan. This quake, called the 2011 Tōhokuearthquake, was the most powerful ever to strike Japan and one of the top five known in the world. Damage from the earthquake was nearly overshadowed by the tsunami it generated, which wiped out coastal cities and towns. Two months after the earthquake, about 25,000 people were dead or missing, and 125,000 buildings had been damaged or destroyed. Aftershocks, some as large as major earthquakes, have continued to rock the region.

The Pacific Northwest of the United States is at risk from a potentially massive earthquake that could strike any time. Subduction of the Juan de Fuca plate beneath

North America produces active volcanoes, but large earthquakes only hit every 300 to 600 years. The last was in 1700, with an estimated magnitude of around 9.0.

Massive earthquakes are the hallmark of the thrust faulting and folding when two continental plates converge. The 2001 Gujarat earthquake in India was responsible for about 20,000 deaths, and many more people became injured or homeless. In Understanding Earthquakes: From Research to Resilience, scientists try to understand the mechanisms that cause earthquakes and tsunamis and the ways that society can deal with them.

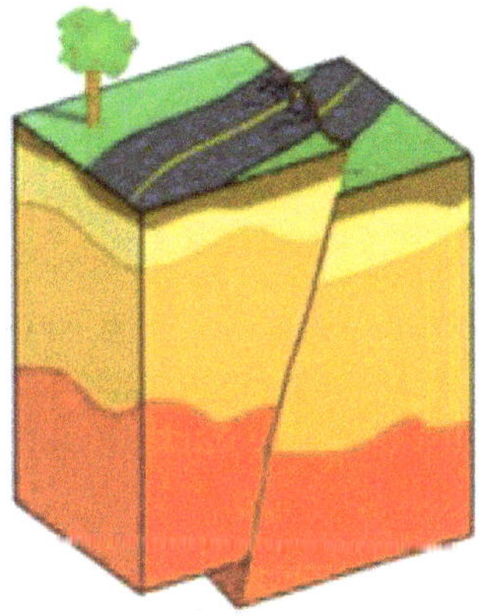

Figure 14.5 Reverse Fault Model. Image is in the public domain.

Divergent Plate Boundary Zone

Earthquakes at mid-ocean ridges are small and shallow because the plates are young, thin, and hot. On land where continents split apart, earthquakes are larger and stronger. A classic example of normal faulting along divergent boundaries is the Wasatch Front in Utah and the entire Basin and Range through Nevada.

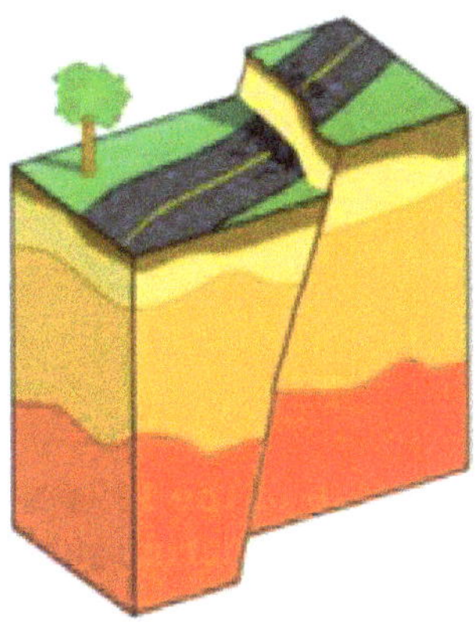

Figure 14.6 Normal Fault Model. Image is in the public domain.

Intraplate Boundary Zone

Intraplate earthquakes are the result of stresses caused by plate motions acting in solid slabs of lithosphere. In 1812, a magnitude 7.5 earthquake struck near New Madrid, Missouri. The earthquake was strongly felt over approximately 50,000 square miles and altered the course of the Mississippi River. Because very few people lived there at the time, only 20 people died.

Many more people live there today. A similar earthquake today would undoubtedly kill many people and cause a great deal of property damage.

Measuring Earthquakes

People have always tried to quantify the size of, and damage done by earthquakes. Since early in the 20th century, there have been three methods. The oldest of the scales is calledthe Mercalli Intensity scale. Earthquakes are described in terms of what nearby residents felt and the damage that was done to nearby structures. This scale is more qualitative in information because it's based on visual damage and not the actual energy released by the earthquake.

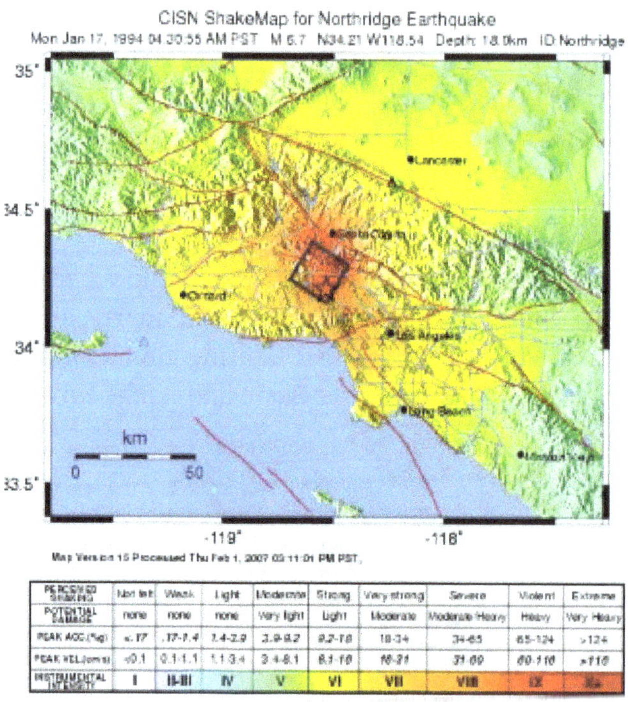

Figure 14.7 Example of a Shake Map of the 1994 Northridge Earthquake. Image used with permission

With the invention of the seismograph station, the Richter magnitude scale was created. Developed in 1935 by Charles Richter, this scale uses a seismometer to measure the magnitude of the largest jolt of energy released by an earthquake.

Today, the moment magnitude scale has replaced the Richter scale. The moment magnitude scale measures the total energy released by an earthquake. The moment magnitude is calculated from the area of the fault that is ruptured and the distance the ground moved along the fault. The Richter scale and the moment magnitude scale are logarithmic. The amplitude of the largest wave increases ten times from one integer to the next. An increase in one integer means that thirty times more energy was released. These two scales often give very similar measurements. How does the amplitude of the largest seismic wave of a magnitude 5 earthquake compare with the largest wave of a magnitude 4 earthquake? How does it compare with a magnitude 3 quake? The amplitude of the largest seismic wave of a magnitude 5 quake is 10 times that of a magnitude 4 quake and 100 times that of a magnitude 3 quake.

Table 14.1 Richter Scale

MAGNITUDE	DESCRIPTION	WHAT IT FEELS LIKE	FREQUENCY
LESS THAN 2.0	Micro	Normally only recorded by seismographs. Most people cannot feel them.	Millions per year.
2.0 – 2.9	Minor	A few people feel them. No building damages.	Over 1 Million per year.
3.0 – 3.9	Minor	Some people feel them. Objects inside can be seen shaking.	Over 100,000 per year.
4.0 – 4.9	Light	Most people feel it. Indoor objects shake or fall to the floor.	10,000 to 15,000 per year.
5.0 – 5.9	Moderate	Can damage or destroy buildings not designed to withstand earthquakes. Everyone feels it.	1,000 to 1,500 per year.
6.0 – 6.9	Strong	Widespread shaking far from the epicenter. Damages buildings.	100 to 150 per year.
7.0 – 7.9	Major	Widespread damage in most areas.	10 to 20 per year.
8.0 – 8.9	Great	Widespread damage in large areas.	About 1 per year.
9.0 – 9.9	Great	Severe damage to most buildings.	1 per 5-50 years.
10.0 OR OVER	Massive	Never recorded.	Never recorded.

Each scale has its own benefits. As mentioned above, the Mercalli Intensity scale is based on how much damage someone would see, or instrumental intensity. This is relative though because some places have strong building codes, and the rock material underneath will impact ground shaking without changing the energy released at the focus. With the Richter scale, a single sharp jolt measures higher than a very long intense earthquake that releases more energy. The moment magnitude scale more accurately reflects the energy released and the damage caused. Today, most seismologists now use the moment magnitude scale.

Earthquake Prediction

Scientists are a long way from being able to predict earthquakes. A good prediction must be accurate as to where an earthquake will occur, when it will occur, and at what magnitude it will be so that people can evacuate. An unnecessary evacuation is expensive and causes people not to believe authorities the next time an evacuation is ordered.

Where an earthquake will occur is the easiest feature to predict. Scientists know that earthquakes take place at plate boundaries and tend to happen where they've occurred before. Earthquake-prone communities should always be prepared for an earthquake. These communities can implement building codes to make structures earthquake safe.

When an earthquake will occur is much more difficult to predict. Since stress on a fault builds up at the same rate over time, earthquakes should occur at regular intervals. But so far scientists cannot predict when quakes will occur even to within a few years. Signs sometimes come before a large earthquake. Small quakes, called foreshocks, sometimes occur a few seconds to a few weeks before a major quake. However, many earthquakes do not have foreshocks and small earthquakes are not necessarily followed by a large earthquake. Often, the rocks around a fault will dilate as microfractures from ground tilting, caused by the buildup of stress in the rocks, may precede a large earthquake, but not always. Water levels in wells fluctuate as water moves into or out of fractures before an earthquake. This is also an uncertain predictor of large earthquakes. The relative arrival times of P-waves and S-waves also decreases just before an earthquake occurs.

Folklore tells of animals behaving erratically just before an earthquake. Mostly these anecdotes are told after the earthquake. If indeed animals sense danger from earthquakes or tsunami, scientists do not know what it is they could be sensing, but they would like to find out.

Damage from Earthquakes

Earthquakes are natural disasters that cause enormous amounts of damage, second only to hurricanes. Earthquake-safe construction techniques, securing heavy objects, and preparing an emergency kit are among the precautions people can take to minimize damage. Earthquakes kill people and cause property damage. However, the ground shaking seldom kills people, and the ground does not swallow someone up. The damage depends somewhat on the earthquake size but mostly on the quality of structures. Structures falling on people injure and kill them.

More damage is done, and more people are killed by the fires that follow an earthquake than the earthquake itself.

Figure 14.8 Photos Above Show Two Low Lying Areas in San Francisco After the 1906 Earthquake, Much Damage Was Contributed to Liquefaction. Image is in the public domain.

There are a variety of ways earthquakes become deadly. Probably the greatest influenceis population density. The magnitude 9.2 Great Alaska Earthquake, near Anchorage, of 1964 resulted in only 131 deaths. At the time few people lived in the area. Oddly enough, size doesn't matter. Only about 2,000 people died in the 1960 Great Chilean earthquake, the largest earthquake ever recorded. The Indian Ocean earthquake of 2004 was one of the largest ever, but most of the 230,000 fatalities were caused by the tsunami, not the earthquake itself.

Ground type is an important factor in earthquake damage. Solid bedrock vibrates less than soft sediments so there is less damage on bedrock. Sediments that are saturated with water undergo liquefaction and become like quicksand.

This will have dangerous implications for Salt Lake City and the Wasatch Front in Utah. Much of the Wasatch Front is loose soil, leftover from the remnants of Lake Bonneville. Along the middle of the two mountain ranges, between the Wasatch Mountains and the Oquirrh Mountains, is the Jordan River which flows from Utah Lake northward to the Great Salt Lake.

Near the river and surrounding floodplain, the water table is near the surface. When the ground shakes, the water near the surface shifts upward and over-saturates the soil making it extremely unstable.

Figure 14.9 Sand Boils Forms Along the Bear River in Utah. Liquefaction is in the public domain.

Human-Induced Earthquakes

Can humans create earthquakes? Maybe not intentionally, but the answer is yes—and here's why. If a water reservoir is built on top of an active fault line, the water may lubricate the fault and weaken the stress built up within it. This may either create a series of small earthquakes or potentially create a large earthquake. Also, the shear weight of the reservoir's water can weaken the bedrock causing it to fracture. Then the obvious concern is if the dam fails.

Earthquakes can also be generated if humans inject other fluids into a fault such as sewage or chemical waste. Finally, nuclear explosions can trigger earthquakes. One way to determine if a nation has tested a nuclear bomb is by monitoring the earthquakes and energy released by the explosion.

SUMMARY

An earthquake is the shaking that results when a body of rock that has been deformed breaks and the two sides quickly slide past each other. The rupture is initiated at a point but quickly spreads across an area of a fault, via a series of aftershocks initiated by stress transfer. Episodic tremor and slip is a periodic slow movement, accompanied by harmonic tremors, along the middle part of a subduction zone boundary.

Most earthquakes take place at or near plate boundaries, especially at transform boundaries (where most quakes are at less than 30 km depth) and at convergent boundaries (where they can be at well over 100 km depth). The largest earthquakes happen at subduction zones, typically in the upper section where the rock is relatively cool.

Magnitude is a measure of the amount of energy released by an earthquake, and it is proportional to the area of the rupture surface and the amount of displacement. Although any earthquake has only one magnitude value, it can be estimated in various ways, mostly involving seismic data. Intensity is a measure of the amount of shaking experienced and damage done at a particular location around the earthquake. Intensity will vary depending on the distance to the epicenter, the depth of the earthquake, and the geological nature of the material below the surface.

Damage to buildings is the most serious consequence of most large earthquakes. The amount of damage is related to the type and size of buildings, how they are constructed, and the nature of the material on which they are built. Other important consequences are fires, damage to bridges and highways, slope failures, liquefaction, and tsunami. Tsunami, which is almost all related to large subduction earthquakes, can be devastating.

15 Chapter

EARTHS DYNAMIC SURFACE: VOLCANOES

Figure 15.1 Geothermal Hot Springs in Hot Creek, Mammoth California. Image by Jeremy Patrich is used under a CC-BY 4.0 license.

Goals & Objectives of this unit

- Explain the relationships between plate tectonics, the formation of magma, and volcanism
- Describe the range of magma compositions formed in differing tectonic environments, and discuss the relationship between magma composition (and gas content) anderuption style
- Explain the geological and eruption-style differences between different types of volcanoes, especially shield volcanoes, composite volcanoes, and cinder cones
- Understand the types of hazards posed to people and infrastructure by the different types of volcanic eruptions

VOLCANISM

A volcano is any location where magma comes to the surface or has done so within the past several million years. This can include eruptions on the ocean floor (or even under the water of lake), where they are called subaqueous eruptions, or on land, where they are called subaerial eruptions. Not all volcanic eruptions produce the volcanic mountains with which we are familiar; in fact, most of Earth's volcanism takes place along the spreading ridges on the seafloor and does not produce volcanic mountains at all, not even sea-floor mountains.

The study of volcanoes is critical to our understanding of the geological evolution of Earth, and our understanding of significant changes in climate. But, most important of all, understanding volcanic eruptions allows us to save lives and property. Over the past few decades, volcanologists have made great strides in their ability to forecast volcanic eruptions and predict the consequences, this has already saved thousands of lives.

Plate Tectonics & Volcanism

Magma is formed at three main plate-tectonic settings: divergent boundaries (decompression melting), convergent boundaries (flux melting), and mantle plumes (decompression melting). Composite volcanoes form at subduction zones, either on ocean-ocean convergent boundaries or ocean-continent convergent boundaries. Both shield volcanoes and cinder cones form in areas of continental rifting. Shield volcanoes form above mantle plumes but can also form at other tectonic settings. Sea-floor volcanism can take place at divergent boundaries, mantle plumes, and ocean-ocean-convergent boundaries.

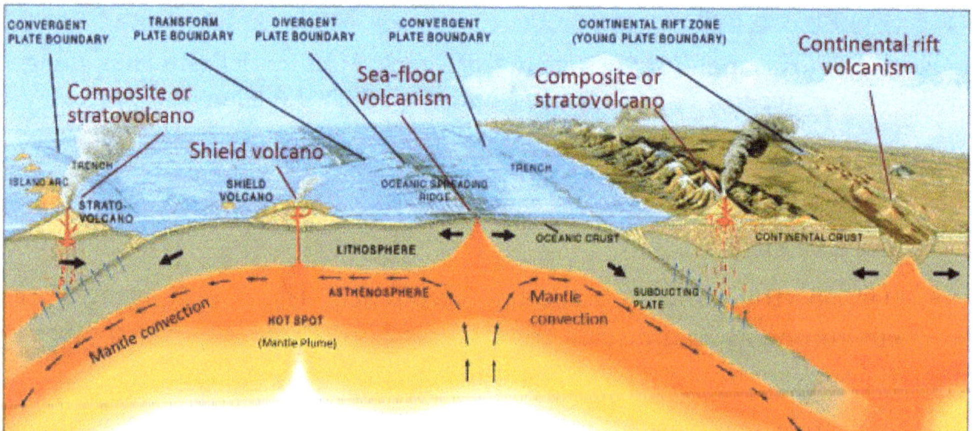

Figure 15.2 The Plate-Tectonic Settings of Common Volcanism. Image by USGS is in the public domain.

The mantle and crustal processes that take place in areas of volcanism. At a spreading ridge, hot mantle rock moves slowly upward by convection (cm/

year), and within about 60 km of the surface, partial melting starts because of decompression. Nearly 10% of the ultramafic mantle rock melts, producing mafic magma that moves upward toward the axis of spreading (where the two plates are moving away from each other). The magma fills vertical fractures produced by the spreading and spills out onto the seafloor to form basaltic pillows (more on that later) and lava flows.

At an ocean-continent or ocean-ocean convergent boundary, oceanic crust is pushed far down into the mantle. It is heated up, and while there isn't enough heat to melt the subducting crust, there is enough to force the water out of some of its minerals. This water rises into the overlying mantle where it contributes to the flux melting of the mantle rock. The mafic magma produced rises through the mantle to the base of the crust. There it contributes to the partial melting of crustal rock, and thus it assimilates much more felsic material. That magma, now intermediate in composition, continues to rise and assimilate crustal material; in the upper part of the crust, it accumulates into plutons. From time to time, the magma from the plutons rises toward the surface, leading to volcanic eruptions. Mount Rainier is an example of subduction- related volcanism.

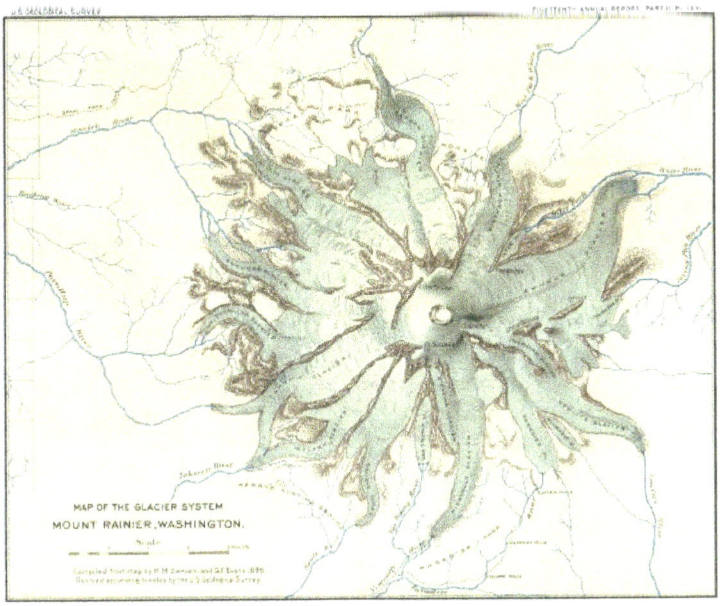

Figure 15.3 Relief Map of Mt. Rainier from 1896. Plate LXVI by USGS is in the public domain.

A mantle plume is an ascending column of hot rock (not magma) that originates deep in the mantle, possibly just above the core-mantle boundary. Mantle plumes are thought to rise at approximately 10 times the rate of mantle convection. The ascending column may be on the order of kilometers to tens of kilometers across, but near the surface, it spreads out to create a mushroom-style head that is several tens to over 100 kilometers across. Near the base of the lithosphere (the rigid part of the mantle), the mantle plume (and possibly some of the surrounding mantle material) partially melt to form mafic magma that rises to feed volcanoes. Since most mantle plumes are beneath the oceans, the early stages of volcanism typically take place on the seafloor. Over time, islands may form like those in Hawaii.

Magma Composition & Eruption Style

As noted in the previous section, the types of magma produced in the various volcanic settings can differ significantly. At divergent boundaries and oceanic mantle plumes, where there is little interaction with crustal materials and magma fractionation to create felsic melts does not take place, the magma tends to be consistently mafic. At subduction zones, where the magma ascends through significant thicknesses of crust, the interaction between the magma and the crustal rock, some of which is quite felsic leads to increases in the felsic character of the magma.

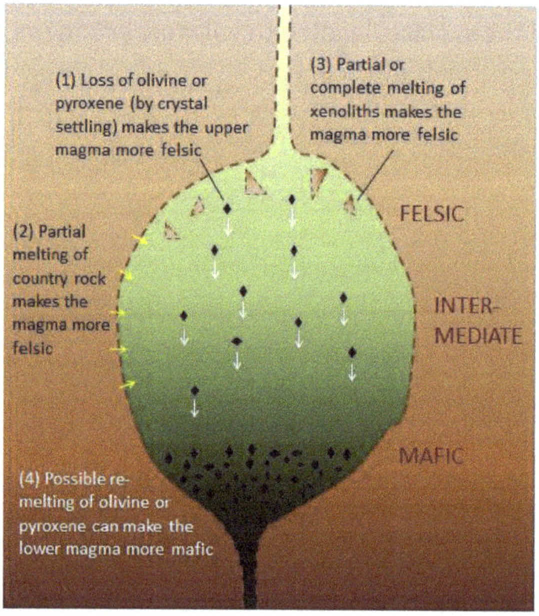

Figure 15.4 Changes in the Composition of Magmas Stored Within a Chamber. Image by Steven Earle, CC BY 4.0.

Several processes can make magma that is stored in a chamber within the crust more felsic and can also contribute to the development of vertical zonation from more mafic at the bottom to more felsic at the top. Partial melting of country rock and country-rock xenoliths increases the overall felsic character of the magma; first, because the country rocks tend to be more felsic than the magma, and second, because the more felsic components of the country-rock melt preferentially. Settling of ferromagnesian crystals from the upper part of the magma, and possible remelting of those crystals in the lower part can both contribute to the vertical zonation from relatively mafic at the bottom to more felsic at the top.

From the perspective of volcanism, there are some important differences between felsic and mafic magmas. First, as we've already discussed, felsic magmas tend to be more viscous because they have more silica, and hence more polymerization. Second, felsic magmas tend to have higher levels of volatiles; that is, components that behave as gases during volcanic eruptions. The most abundant volatile in magma is water (H2O), followed typically by carbon dioxide (CO2), and then by sulphur dioxide (SO2). The general relationship between the SiO2content of magma and the number of volatiles is shown in Figure 4.8. Although there are many exceptions to this trend, mafic magmas typically have 1% to 3% volatiles, intermediate magmas have 3% to 4% volatiles, and felsic magmas have 4% to 7% volatiles.

Differences in viscosity and volatile levels have significant implications for the nature of volcanic eruptions. When magma is deep beneath the surface and under high pressure from the surrounding rocks, the gases remain dissolved. As magma approaches the surface, the pressure exerted on it decreases. Gas bubbles start to form, and the more gas there is in the magma, themore bubbles form. If the gas content is low or the magma is runny enough for gases to rise through it and escape to the surface, the pressure will not become excessive. Assuming that it can break through to the surface, the magma will flow out relatively gently. An eruption that involves a steady non-violent flow of magma is called effusive.

If the magma is felsic, and therefore too viscous for gases to escape easily, or if it has a particularly high gas content, it is likely to be under high pressure. Viscous magma doesn't flow easily, so even if there is a way for it to move out, it may not flow out. Under these circumstances, pressure will continue to build as more magma moves up from beneath and gases continue to exsolve. Eventually, some part of the volcano will break and then all of that pent-up pressure will lead to an explosive eruption.

Mantle plume and spreading-ridge magmas tend to be consistently mafic, so effusive eruptions are the norm. At subduction zones, the average magma composition is likely to be close to intermediate, but as we've seen, magma chambers can become zoned and so compositions ranging from felsic to mafic are possible. Eruption styles can be correspondingly variable.

TYPES OF VOLCANOES

The sizes and shapes of a typical shield, composite, and cinder cone volcanoes are compared in the table below, although, to be fair, Mauna Loa is the largest shield volcano on Earth; all others are smaller. Mauna Loa rises from the surrounding flat seafloor, and its diameter is in the order of 200 km. Its elevation is 4,169 m above sea level. Mt. St. Helens, a composite volcano, rises above the surrounding hills of the Cascade Range. Its diameter is about 6 km, and its height is 2,550 m above sea level. Cinder cones are much smaller. On this drawing, even a large cinder cone is just a dot.

Table 15.1 Types of Volcanoesi

Type	Tectonic Setting	Size and Shape	Magna and Eruption Characteristics	Example
Cinder Cone	Various; some form on the flanks of larger volcanoes.	Small: 10 to 100's of meters. Steep: (>20°)	Most are mafic and form from the gas-rich early stages of a shield – or rift associated eruption.	Mt. Lassen
Composite Volcano	Almost all are at subduction zones.	Medium: 1000's of meters. Moderate steepness (10° to 30°)	Magma composition varies from felsic to mafic, and from explosive to effusive.	Mt. St. Helens
Shield Volcano	Most are at mantle plumes; some are on spreading ridges.	Large:1,000 m high and 200 km across. Not steep (2° to 10°)	Magna is almost always mafic, and eruptions are typically effusive, although cinder cones are common on the flanks of shield volcanoes.	Kilauea Hawaii
Large Igneous Provinces	Associated with "super" mantle plumes	Enormous: Millions of km2 and 100s of m thick.	Magma is always mafic and individual flows can be 10s of meters thick.	Columbia River Basalts
Seafloor Volcanism	Generally associated with spreading ridges but also with mantle plumes.	Large areas of seafloor associated with spreading ridges.	A typical eruption rates, pillows form; at faster rates, lava flows develop.	Juan de Fuca Ridge

Cinder Cones

Cinder cones, like Red Mountain in the Eastern Sierra, are typically only a few hundred meters in diameter, and few are more than 200 m high. Most are made up of fragments of vesicular mafic rock (scoria) that were expelled as the magma boiled when it approached the surface, creating fire fountains. In many cases, these later became effusive (lava flows) when the gases were depleted. Most cinder cones are monogenetic, meaning that they formed during a singleeruptive phase that might have lasted weeks or months. Because cinder cones are made up almost exclusively of loose fragments, they have very little strength. They can be easily, and relatively quickly, eroded.

Composite

Composite volcanoes, like Mt. St. Helens in Washington State, are almost all associated with subduction at convergent plate boundaries — either ocean-continent or ocean-ocean boundaries. They can extend up to several thousand meters from the surrounding terrain, and, with slopes ranging up to 30°, are typically up to 10 km across. At many such volcanoes, magma is stored in a magma chamber in the upper part of the crust. For example, at Mt. St.

Figure 15.5 Columnar Basalt at The Devils Postpile Near Mammoth, California. Image by Jeremy Patrich is used under a CC-BY 4.0 license.

Helens, there is evidence of a magma chamber that is approximately 1 km wide and extends from about 6 km to 14 km below the surface. Systematic variations in the composition of volcanism over the past several thousand years at Mt. St. Helens imply that the magma chamber is zoned, from more felsic at the top to more mafic at the bottom.

Mafic eruptions (and some intermediate eruptions), on the other hand, produce lava flows; the one shown below is thick enough (about 10 m in total) to have cooled in a columnarjointing pattern. Lava flows both flatten the profile of the volcano (because the lava typically flows farther than pyroclastic debris falls) and protect the fragmental deposits from erosion. Even so, composite volcanoes tend to erode quickly. Patrick Pringle, a volcanologist with the Washington State Department of Natural Resources, describes Mt. St. Helens as a "pilc of junk."

The rock that makes up Mt. St. Helens ranges in composition from rhyolite to basalt; this implies that the types of past eruptions have varied widely in character. As already noted, felsic magma doesn't flow easily and doesn't allow gases to escape easily. Under these circumstances, pressure builds up until a conduit opens, and then an explosive eruption results from the gas-rich upper part of the magma chamber, producing pyroclastic debris. This type of eruption can also lead to the rapid melting of ice and snow on a volcano, which typically triggers large mudflows known as lahars. Hot, fast-moving pyroclastic flows and lahars are the two main causes of casualties in volcanic eruptions. Pyroclastic flows killed approximately 30,000 people during the 1902 eruption of Mt. Pelée on the Caribbean island of Martinique.

Most were incinerated in their homes. In 1985 a massive lahar, triggered by the eruption of Nevado del Ruiz, killed 23,000 people in the Colombian town of Armero, about 50 km from the volcano.

In a geological context, composite volcanoes tend to form relatively quickly and do not last very long. Mt. St. Helens, for example, is made up of rock that is all younger than 40,000 years; most of it is younger than 3,000 years. If its volcanic activity ceases, it might erode within a few tens of thousands of years. This is largely because of the presence of pyroclastic eruptive material, which is not strong.

Shield

Most shield volcanoes are associated with mantle plumes, although some form at divergent boundaries, either on land or on the seafloor. Because of their non-viscous mafic magma, they tend to have relatively gentle slopes (2 to 10°) and the larger ones can be over 100 km in diameter. The best-known shield volcanoes are those that make up the Hawaiian Islands, and of these, the only active ones are on the big island of Hawaii. Mauna Loa, the world's largest volcano and the world's largest mountain (by volume) last erupted in 1984. Kilauea, arguably the world's most active volcano, has been erupting, virtually without interruption, since 1983.

Loihi is an underwater volcano on the southeastern side of Hawaii. It is last known to have erupted in 1996 but may have erupted since then without being detected.

All of the Hawaiian volcanoes are related to the mantle plume that currently lies beneath Mauna Loa, Kilauea, and Loihi. In this area, the Pacific Plate is moving northwest at a rate of about 7 cm/year. This means that the earlier formed, and now extinct, volcanoes have now moved well away from the mantle plume. There is evidence of crustal magma chambers beneath all three active Hawaiian volcanoes. At Kilauea, the magma chamber appears to be several kilometers in diameter and is situated between 8 km and 11 km below the surface Although it is not a prominent mountain, Kilauea volcano has a large caldera in its summit area. A caldera is a volcanic crater that is more than 2 km in diameter; this one is 4 km long and 3 kmwide. It contains a smaller feature called Halema'uma'u crater, which has a total depth of over 200 m below the surrounding area. Most volcanic craters and calderas are formed above magma chambers, and the level of the crater floor is influenced by the amount of pressure exerted by the magma body. During historical times, the floors of both Kilauea caldera and Halema'uma'u crater have moved up during the expansion of the magma chamber and down during deflation of the chamber.

The two main types of textures created during effusive subaerial eruptions are pahoehoe and aa. Pahoehoe, ropy lava that forms as non-viscous lava, flows gently, forming a skin that gels and then wrinkles because of the ongoing flow of the lava below the surface.

Aa, or blocky lava, forms when magma is forced to flow faster than it can (down a slope for example). Tephra, lava fragments, is produced during explosive eruptions and accumulates in the vicinity of cinder cones.

Below is a view of an extinct lava tube in Mojave, California. Very thick, gooey rhyolitic lava doesn't flow very far. The runny basaltic sort that characterizes some of the lava flows of Mojave National Preserve, however, spreads out as smoothly as hot maple syrup. It flowed from the sides of the cones or pooled near their bases. As the lava streamed out across the land, it slowly began to cool. Often, the top of flow would cool while liquid lava continued moving underneath, creating a tunnel. When the eruption ended, the flowing lava in the tunnel either cooled in place or emptied the tunnel's end, leaving a hollow lava tube. A lava tube is accessible via a 5-mile drive from Kelbaker Road. Climbing through a collapsed hole in the tube's roof, visitors have a rare opportunity to view this river of rock from the lava's perspective.

Figure 15.6 Photo from Inside the Mojave Volcanic Tubes. Image by Jeremy Patrich is used under a CC-BY 4.0 license.

Sea-Floor Volcanism

Some large igneous provinces eruptions occur on the seafloor, the largest being the one that created the Ontong Java Plateau in the western Pacific Ocean at around 122 Ma. But most sea- floor volcanism originates at divergent boundaries and involves relatively low-volume eruptions. Under these conditions, hot lava that oozes out into the cold seawater quickly cools on the outside and then behaves a little like toothpaste. The resulting blobs of lava are known as pillows, and they tend to form piles around a sea-floor lava vent. In terms of area, there is very likely more pillow basalt on the seafloor than any other type of rock on Earth.

VOLCANIC HAZARDS

There are two classes of volcanic hazards, direct and indirect. Direct hazards are forces that directly kill or injure people or destroy property or wildlife habitat. Indirect hazards are volcanism-induced environmental changes that lead to distress, famine, or habitat destruction. Indirect effects of volcanism have accounted for approximately 8 million deaths during historical times, while direct effects have accounted for fewer than 200,000, or 2.5% of the total.

Volcanic Gas & Tephra Emissions

Large volumes of tephra (rock fragments, mostly pumice) and gases are emitted duringmajor Plinian eruptions (large explosive eruptions with hot gas a tephra column extending into the stratosphere) at composite volcanoes, and a large volume of gas is released during some very high-volume effusive eruptions. One of the major effects is cooling of the climate by 1° to 2°C for several months to a few years because the dust particles and tiny droplets and particles of sulfur compounds block the sun. The last significant event of this type was in 1991 and 1992 following the large eruption of Mt. Pinatubo in the Philippines. A drop of 1° to 2°C may not seem like very much, but that is the global average amount of cooling, and cooling was much more severe in some regions.

Over eight months in 1783 and 1784, a massive effusive eruption took place at the Laki volcano in Iceland. Although there was relatively little volcanic ash involved, a massive amount of sulfur dioxide was released into the atmosphere, along with a significant volume of hydrofluoric acid (HF). The sulphate aerosols that formed in the atmosphere led to dramatic cooling in the Northern Hemisphere. There were serious crop failures in Europe and North America, and a total of 6 million people are estimated to have died from famine and respiratory complications. In Iceland, poisoning from the HF resulted in the death of 80% of sheep, 50% of cattle, and the ensuing famine, along with HF poisoning, resulted in more than 10,000 human deaths, about 25% of the population.

Volcanic ash can also have serious implications for aircraft because it can destroy jet engines. For example, over 5 million airline passengers had their travel disrupted by the 2010 Eyjafjallajökull volcanic eruption in Iceland.

Pyroclastic Density Currents

In a typical explosive eruption at a composite volcano, the tephra and gases are ejected with explosive force and are hot enough to be forced high up into the atmosphere. As the eruption proceeds and the amount of gas in the rising magma starts to decrease, parts will become heavier than air, and they can then flow downward along the flanks of the volcano. As they descend, they cool more and flow faster, reaching speeds up to several hundred km/h.

A pyroclastic density current (PDC) consists of tephra ranging in size from boulders to microscopic shards of glass (made up of the edges and junctions of the bubbles of shattered pumice), plus gases (dominated by water vapor, but also including other gases). The temperature of this material can be as high as 1000°C. Among the most famous PDCs are the one that destroyed Pompeii in the year 79 CE, killing an estimated 18,000 people, and the one that destroyed the town of St. Pierre, Martinique, in 1902, killing an estimated 30,000.

Figure 15.7 An Example of Pyroclastic Debris, Mt. Saint Helens 1980. Image by Dr. Bob Burnstein, MD, Ph.D., MPH is in the public domain.

Pyroclastic Fall

Most of the tephra from an explosive eruption ascends high into the atmosphere, and some of it is distributed around Earth by high-altitude winds. The larger components (larger than 0.1 mm) tend to fall relatively close to the volcano, and the amount produced by large eruptions can cause serious damage and casualties. The large 1991 eruption of Mt. Pinatubo in the Philippines resulted in the accumulation of tens of centimeters of ash in fields and on rooftops in the surrounding populated region. Heavy typhoon rains that hit the island at the same time added to the weight of the tephra, leading to the collapse of thousands of roofs and at least 300 of the 700 deaths attributed to the eruption.

Lahar

A lahar is any mudflow or debris flow that is related to a volcano. Most are caused by melting snow and ice during an eruption, as was the case with the lahar that destroyed the Colombian town of Armero in 1985 (described earlier). Lahars can also happen when there is no volcanic eruption, and one of the reasons is that, as we've seen, composite volcanoes tend to be weak and easily eroded.

In October 1998, category 5 Hurricane Mitch slammed into the coast of Central America. The damage was extensive and 19,000 people died, not so much because of high winds but becauseof intense rainfall, some regions received almost 2 m of rain over a few days! Mudflows and debris flow occurred in many areas, especially in Honduras and Nicaragua. An example is Casita Volcano in Nicaragua, where the heavy rains weakened rock and volcanic debris on the upper slopes, resulting in a debris flow that rapidly built-in volume as it raced down the steep slope, and then ripped through the towns of El Porvenir and Rolando Rodriguez killing more than 2,000 people. El Porvenir and Rolando Rodriguez were new towns that had been built without planning approval in an area that was known to be at risk of lahars.

SUMMARY

Volcanism is closely related to plate tectonics. Most volcanoes are associated with convergent plate boundaries (at subduction zones), and there is also a great deal of volcanic activity at divergent boundaries and areas of continental rifting. At convergent boundaries magma is formed where water from a subducting plate acts as a flux to lower the melting temperature of the adjacent mantle rock. At divergent boundaries magma forms because of decompression melting. Decompression melting also takes place within a mantle plume.

The initial magmas in most volcanic regions are mafic in composition, but they can evolve into more felsic types through interaction with crustal rock, and as a result of crystal settling within a magma chamber. Felsic magmas tend to have higher gas contents than mafic magmas, and they are also more viscous. The higher viscosity prevents gases from escaping from the magma, and so felsic magmas are more pressurized and more likely to erupt explosively.

Cinder cones, which can form in various volcanic settings, are relatively small volcanoes that are composed mostly of mafic rock fragments that were formed during a single eruptiveevent. Composite volcanoes are normally associated with subduction, and while their magma tends to be intermediate on average, it can range from felsic to mafic. The corresponding differences in magma viscosity lead to significant differences in eruptions style. Most shield volcanoes are associated with mantle plumes and have consistently mafic magma which generally erupts as lava flows.

Most direct volcanic hazards are related to volcanoes that erupt explosively, especially composite volcanoes. Pyroclastic density currents, some as hot as 1000°C can move at hundreds of km/h and will kill anything in the way. Lahars, volcano-related mudflows, can belarge enough to destroy entire towns. Lava flows will destroy anything in their paths but tend to move slowly enough so that people can get to safety.

Chapter 16 SHAPED BY COASTAL PROCESSES

Figure 16.1 Coral Beach in Cancun. Image by Jeremy Patrich is used under a CC-BY 4.0 license.

Goals & Objectives of this unit

- Describe the origins of longshore currents and longshore drift
- Explain why some coasts are more affected by erosion than others and describe the formation of coastal erosional features, including stacks, arches, cliffs, and wave-cut platforms
- Summarize the origins of beaches, spits, baymouth bars, tombolos, and barrier islands

- Explain the various mechanisms of sea-level change and the implications for coastal processes
- Compare the positive and negative implications of human interference with coastalprocesses

SHORELINES

Most people love shorelines. They love panoramic ocean views, they love sandy beaches on crystal-clear lakes, they love to swim and surf and go out in boats, and they love watching the giant waves crash onto rocky shores. While an understanding of coastal processes isn't necessary for our enjoyment of coastal regions, it can make our time there much more interesting. But an understanding of coastal processes is critical to people who live near a coast, or those who like to spend a lot of time there, because, to be safe and avoid damage to infrastructure, we need to know how coastal processes work. We also need to understand the processes in order to avoid some of the possible consequences of changes that we might like to make in coastal areas.

Waves

Waves form on the ocean and lakes because energy from the wind is transferred to the water. The stronger the wind, the longer it blows, and the larger the area of water over which it blows (the fetch), the larger the waves are likely to be.

The important parameters of a wave are its wavelength (the horizontal distance between two crests or two troughs), its amplitude (the vertical distance between a trough and a crest), and its velocity (the speed at which wave crests move across the water)

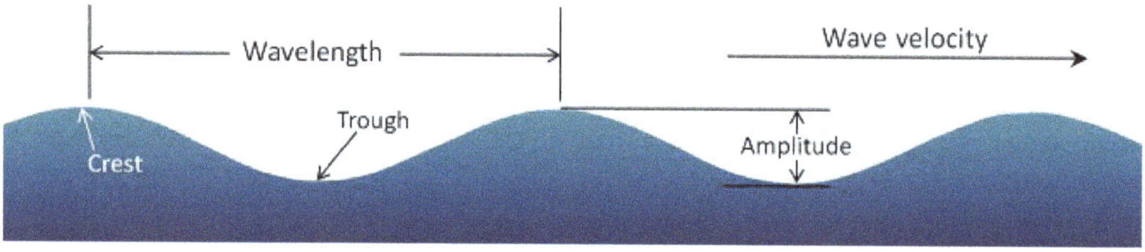

Figure 16.2 The Parameters of Water Waves. Image by Steven Earle, CC BY 4.0.

The typical sizes and speeds of waves in situations where they have had long enough to develop fully are summarized in the table below. In a situation where the fetch is short (say 19 km on a lake) and the wind is only moderate (19 km/h), the waves will develop fully within 2 hours, but they will remain quite small (average amplitude about 27 cm, wavelength 8.5 m). On a large body of water (the ocean or a very large lake) with a fetch of 139 km and winds of 37 km/h, the waves will develop fully in 10 hours; the average amplitude will be around 1.5 m and the average wavelength around 34 m. In the open ocean, with strong winds (92 km/h) that blow for at least 69 hours, the waves will average nearly 15 m high and their wavelengths will be over 200 m. Small waves (amplitudes under a meter) tend to have relatively shallow slopes (amplitude is 3% to 4% of wavelength), while larger waves (amplitudes over 10 m) have much steeper slopes (amplitude is 6% to 7% of wavelength). In other words, not only are large waves bigger than small ones they are also generally more than twice as steep, and therefore many times more impressive. It is important to recognize, however, that amplitudes decrease with distance from the area where the waves were generated.

Table 16.1 The Duration times listed are the Minimum Required for Waves to Develop Fully[ii]

Wind Speed	Fetch	Duration	Amplitude	Wavelength	Wave period	Wave Velocity	
km/h	km	h	m	m	s	m/s	km/h
19	19	2	.27	8.5	3.0	2.8	10.2
37	139	10	1.5	33.8	5.7	5.9	19.5
56	518	23	4.1	76.5	8.6	8.9	32.0
74	1,313	42	8.5	136	11.4	11.9	42.9
92	2,627	69	14.8	212	14.3	14.8	53.4

Relatively small waves move at up to about 10 km/h and arrive on a shore about once every 3 seconds. Very large waves move about five times faster (over 50 km/h), but because their wavelengths are so much longer, they arrive less frequently – about once every 14 seconds.

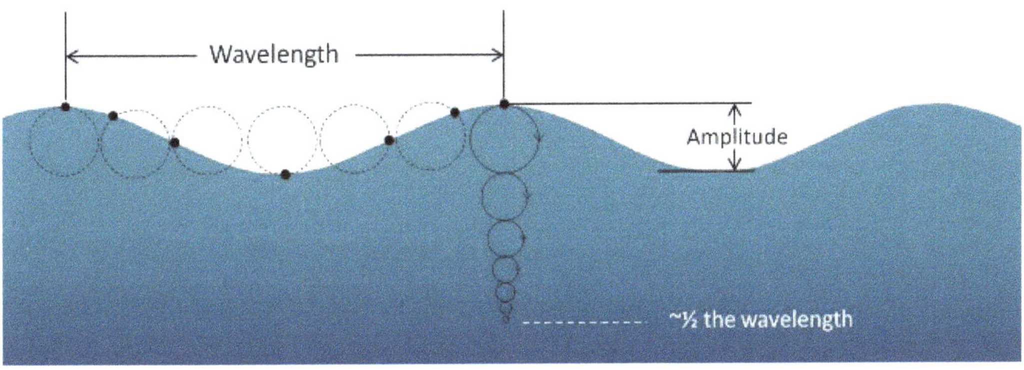

Figure 16.3 The Orbital Motion of a Parcel of Water (Black Dot) as a Wave Moves across the Surface (Image by Steven Earle, CC BY 4.0).

As a wave moves across the surface of the water, the water itself mostly just moves up and down and only moves a small amount in the direction of wave motion. As this happens, a point on the water surface describes a circle with a diameter that is equal to the wave amplitude. This motion is also transmitted to the water underneath, and the water is disturbed by a wave to a depth of approximately one-half of the wavelength.

The one-half wavelength depth of disturbance of the water beneath a wave is known asthe wave base. Since ocean waves rarely have wavelengths greater than 200 m, and the openocean is several thousand meters deep, the wave base does not normally interact with the bottom of the ocean. However, as waves approach the much shallower water near the shore, they start to feel the bottom, and they are affected by that interaction. The wave "orbits" are both flattened and slowed by dragging, and the implications are that the wave amplitude (height) increases and, the wavelength decreases (the waves become much steeper). The ultimate result of this is that the waves lean forward, and eventually break.

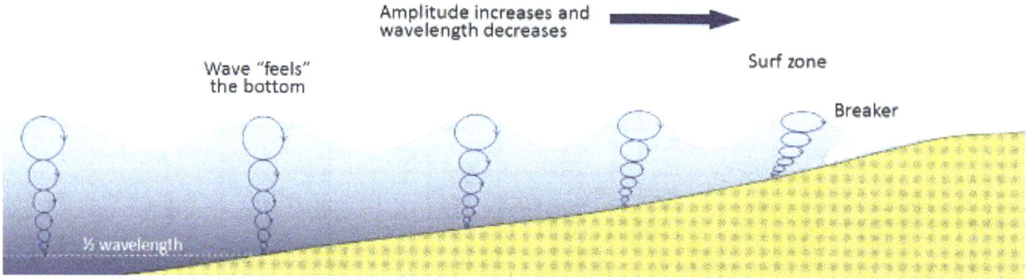

Figure 16.4 The Effect of Waves Approaching a Sandy Shore. Image by Steven Earle, CC BY 4.0.

Waves normally approach the shore at an angle, and this means that one part of the wave feels the bottom sooner than the rest of it, so the part that feels the bottom first slows down first.

Figure 16.6 As Waves Approach the Shore, They are refracted to Become More Parallel to the Beach. Image by Steven Earle, CC BY 4.0.

The waves (with crests shown as white lines in the diagram) were approaching at an angle of about 20° to the beach. The waves first reached shore at the southern end (right side of the image). As they moved into shallow water, they were slowed more at the southern end, and thus gradually became more parallel to the beach.

In open water, these waves had wavelengths close to 100 m. In the shallow water closer to shore, the wavelengths decreased to around 50 m, and in some cases, even less.

Even though they bend and become nearly parallel to the shore, most waves still reach the shore at a small angle, and as each one arrives, it pushes water along the shore, creating what is known as a longshore current within the surf zone, or the areas where waves are breaking.

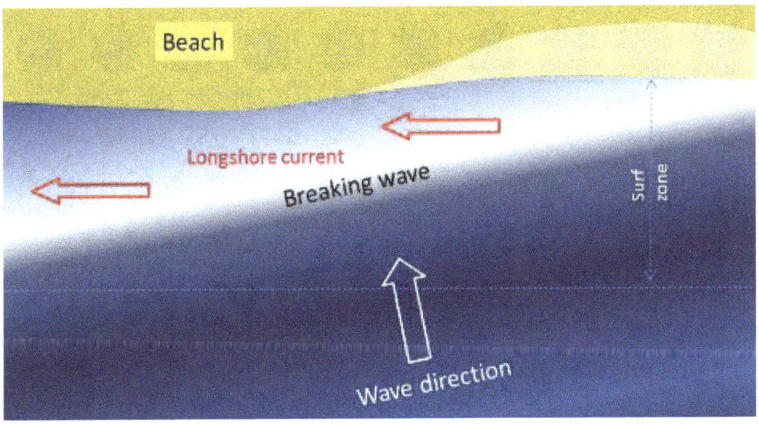

Figure 16.5 The Generation of a Longshore Current by Waves approaching the Shore at an Angle (Image by Steven Earle, CC BY 4.0.

Another important effect of waves reaching the shore at an angle is that when they wash up onto the beach, they do so at an angle, but when that same wave water washes back down the beach, it moves straight down the slope of the beach. The upward-moving water, known as the swash, pushes sediment particles along the beach, while the downward-moving water,the backwash, brings them straight back. With every wave that washes up and then down the beach, particles of sediment are moved along the beach in a zigzag pattern.

The combined effects of sediment transport within the surf zone by the longshore current and sediment movement along the beach by swash and backwash is known as longshore drift.Longshore drift moves a tremendous amount of sediment along coasts (both oceans and large lakes) around the world.

Tides are related to very long-wavelength but low-amplitude waves on the ocean surface (and to a much lesser extent on very large lakes) that are caused by variations in the gravitational effects of the Sun and Moon. Tide amplitudes in shoreline areas vary quite dramatically from place to place. On the west coast of Canada, the tidal range is relatively high, in some areas as much as 6 m, while on most of the east coast the range is lower, typically around 2 m. A major exception is the Bay of Fundy between Nova Scotia and New Brunswick, where the daily range can be as great as 16 m. Anomalous tides like that are related to the shape and size of bays and inlets, which can significantly enhance the amplitude of the tidal surge. The Bay of Fundy has a natural oscillation cycle of 12.5 hours, and that matches the frequency of the rise and fall of the tides in the adjacent Atlantic Ocean. Ungava Bay, on Quebec's north coast, has a similarly high tidal range.

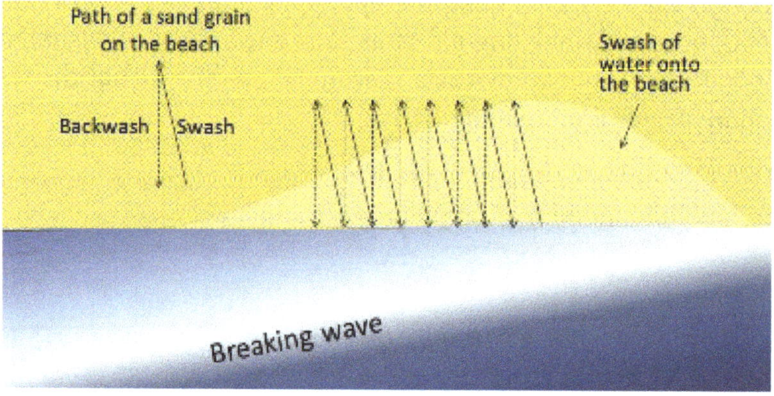

Figure 16.7 The Movement of Particles on a Beach as a Result of Swash and Backwash. Image by Steven Earle, CC BY 4.0.

A rip current is another type of current that develops in the nearshore area and has the effect of returning water that has been pushed up to the shore by incoming waves. As shown in Figure 16.8, rip currents flow straight out from the shore and are fed by the longshore currents. They die out quickly just outside the surf zone but can be dangerous to swimmers who get caught in them. If part of a beach does not have a strong unidirectional longshore current, the rip currents may be fed by longshore currents going in both directions.

Figure 16.8 Rip Currents on Tunquen Beach in Central Chile. Photo by NOAA is in the public domain.

As the tides rise and fall, they push and pull a large volume of water in and out of bays and inlets and around islands. They do not have as significant an impact on coastal erosion anddeposition as wind waves do, but they have an important influence on the formation of features within the intertidal zone.

LANDFORMS OF COASTAL EROSION & DEPOSITION

Large waves crashing onto a shore bring a tremendous amount of energy that has a significant eroding & depositional effect, and several unique erosion & depositional features form on rocky & sandy shores.

Erosional Landforms

When waves approach an irregular shore, they are slowed down to varying degrees, depending on differences in the water depth, and as they slow, they are bent or refracted. That energy is evenly spaced out in the deep water, but because of refraction, the energy of the waves — which moves perpendicular to the wave crests — is being focused on the headlands. On irregular coasts, the headlands receive much more wave energy than the intervening bays, and thus they are more strongly eroded. The result of this is coastal straightening.

Figure 16.9 An Arch in Tilted Sedimentary Rock at the Mouth of the Barachois River, Newfoundland, July 2012. Bottom: The Same Location in June 2013, Showing that the Arch Has Collapsed. Photo by David Murphy in Physical Geology by Steven Earle, CC BY 4.0.

Arches and sea caves are related to stacks because they all form as a result of the erosion of relatively non-resistant rock. An arch in the Barachois River area of western Newfoundland is shown below. This feature started as a sea cave, and then, after being eroded from both sides, became an arch. During the winter of 2012/2013, the arch collapsed, leaving a small stack at the end of the point. If you look carefully at the upper photograph you can see that the hole that makes the arch developed within a layer of relatively soft and weak rock

Depositional Landforms

Some coastal areas are dominated by erosion, an example being the Pacific coast of Canada and the United States, while others are dominated by deposition, examples being the Atlantic and Caribbean coasts of the United States. But on almost all coasts, both deposition and erosion are happening to varying degrees, most of the time, although in different places. This is evident in the Malibu area where erosion is the predominant process on the sedimentary headlands, while depositional processes predominate within the bays. On deposition-dominant coasts, the coastal sediments are still being eroded from some areas and deposited in others.

The main factor in determining if a coast is dominated by erosion or deposition is its history of tectonic activity. The coasts of the United States along the Atlantic and the Gulf of Mexico have not seen significant tectonic activity in a few hundred million years, and except in the northeast, have not experienced post-glacial uplift. These areas have relatively little topographic relief, and there is now minimal erosion of coastal bedrock.

On coasts that are dominated by depositional processes, most of the sediment being deposited typically comes from large rivers. An obvious example is where the Mississippi River flows into the Gulf of Mexico at New Orleans; another being the Los Angeles River flowing out to the Pacific.

On a sandy marine beach, the beach face is the area between the low and high tide levels.A berm is a flatter region beyond the reach of high tides; this area stays dry except during large storms.

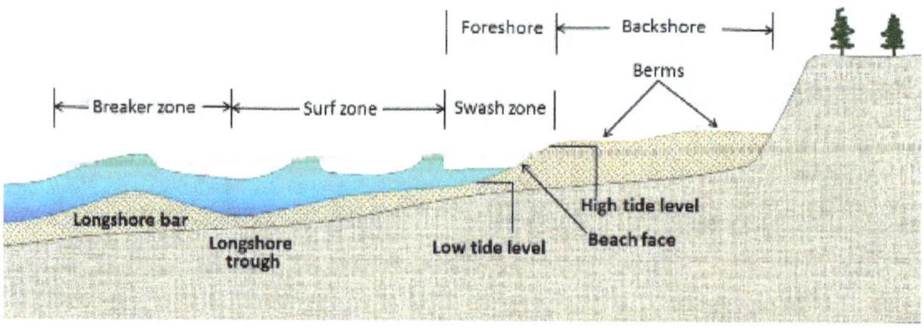

Figure 16.10 The Components of a Sandy Marine Beach. Image by Steven Earle, CC BY 4.0.

Most beaches go through a seasonal cycle because conditions change from summer to winter. In summer, sea conditions are relatively calm with long-wavelength, low-amplitude waves generated by distant winds. Winter conditions are rougher, with shorter-wavelength, higher- amplitude waves caused by strong local winds. The heavy seas of winter gradually erode sand from beaches, moving it to an underwater sandbar offshore from the beach. The gentler waves of summer gradually push this sand back toward the shore, creating a wider and flatter beach.

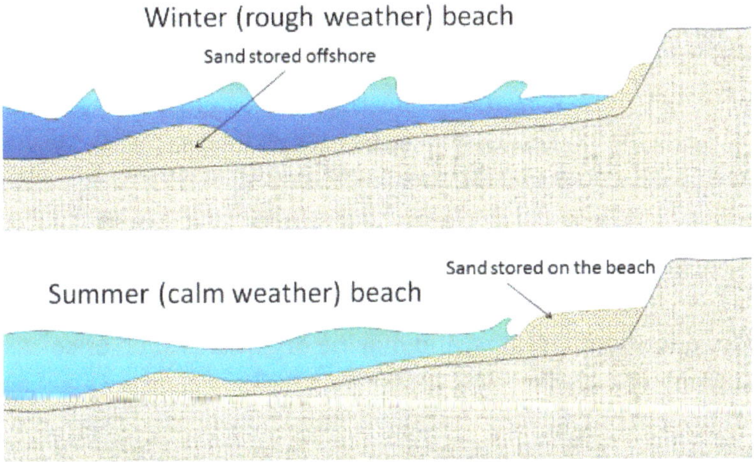

Figure 16.11 The Differences Between Summer & Winter on Beaches in Areas Where the Winter Conditions are Rougher & Waves Have a Shorter Wavelength but Higher Energy. In winter, Sand from the Beach is Stored Offshore in Sand Bars. Image by Steven Earle, CC BY 4.0.

A spit, for example, is an elongated sandy deposit that extends out into open water in the direction of a longshore current.

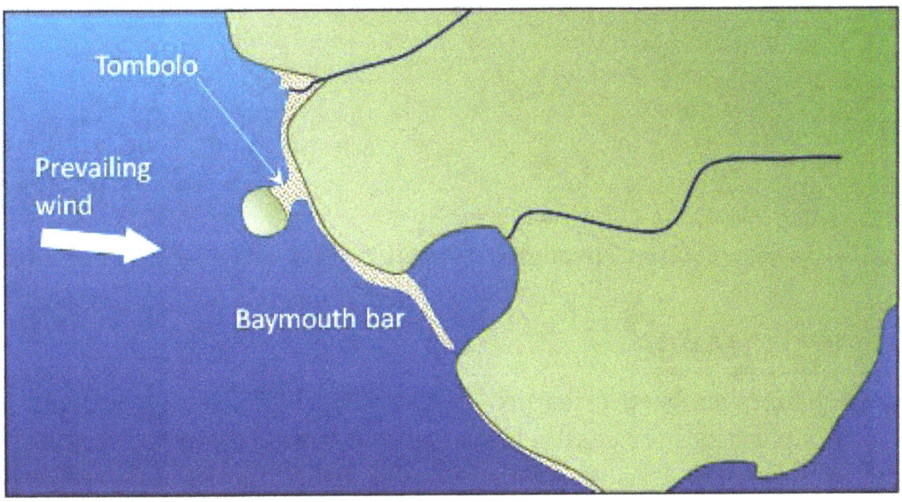

Figure 16.12 A Depiction of a Baymouth Bar & a Tombolo. Image by Steven Earle, CC BY 4.0.

A spit that extends across a bay to the extent of closing, or almost closing it off, is known asa baymouth bar. Most bays have streams flowing into them, and since this water has to get out, rarely a baymouth bar will completely close the entrance to a bay. In areas where there is sufficient sediment being transported, and there are near-shore islands, a tombolo may form.

In areas where coastal sediments are abundant and coastal relief is low (because there has been little or no recent coastal uplift), it is common for barrier islands to form. Barrier islands are elongated islands composed of sand that form a few kilometers away from the mainland. They are common along the U.S. Gulf Coast from Texas to Florida, and along the U.S. Atlantic Coast from Florida to Massachusetts. North of Boston, the coast becomes rocky, partly because that area has been affected by a post-glacial crustal rebound.

Some coasts in tropical regions (between 30° S and 30° N) are characterized by carbonate reefs. Reefs form in relatively shallow marine water within a few hundred to a few thousand meters of shore in areas where there is little or no input of clastic sediments from streams, and marine organisms such as corals, algae, and shelled organisms can thrive. The associated biological processes are enhanced where upwelling currents bring chemical nutrients from deeper water (but not so deep that the water is cooler than about 25°C). Sediments that form in the back reef (shore side) and fore reef (ocean side) are typically dominated by carbonate fragments eroded from the reef and from organisms that thrive in the back-reef area that is protected from wave energy by the reef.

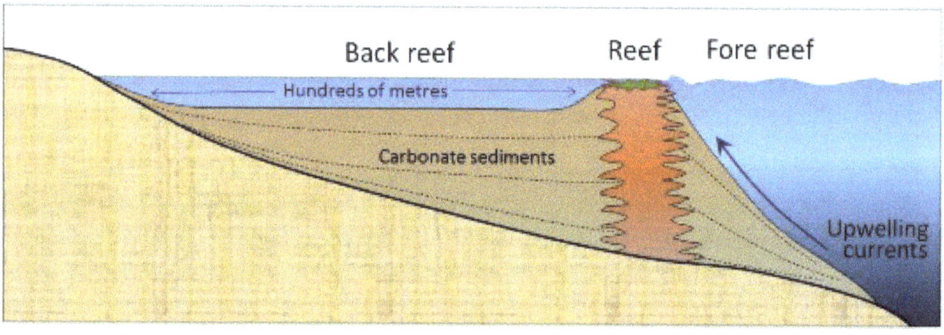

Figure 16.13 Cross-Section Through a Typical Barrier or Fringing Reef. Image by Steven Earle, CC BY 4.0.

SEA-LEVEL CHANGE

Sea-level change has been a feature on Earth for billions of years, and it has important implications for coastal processes and both erosional and depositional features. There are three main mechanisms of sea-level change, as described below.

Eustatic sea-level changes are global sea-level changes related either to changes in the volume of glacial ice on land or to changes in the shape of the sea floor caused by plate tectonic processes. For example, changes in the rate of mid-ocean spreading will change the shape of the seafloor near the ridges, and this affects sea-level.

Over the past 20,000 years, there have been approximately 125 m of eustatic sea-level rise due to glacial melting. Most of that took place between 15,000 and 7,500 years ago during the major melting phase of the North American and Eurasian Ice Sheets. At around 7,500 years ago, the rate of glacial melting and sea-level rise decreased dramatically, and since that time, the average rate has been in the order of 0.7 mm/year. Anthropogenic climate change led to an accelerating sea-level rise starting around 1870. Since that time, the average rate has been 1.1 mm/year, but it has been gradually increasing. Since 1992, the average rate has been 3.2 mm/year.

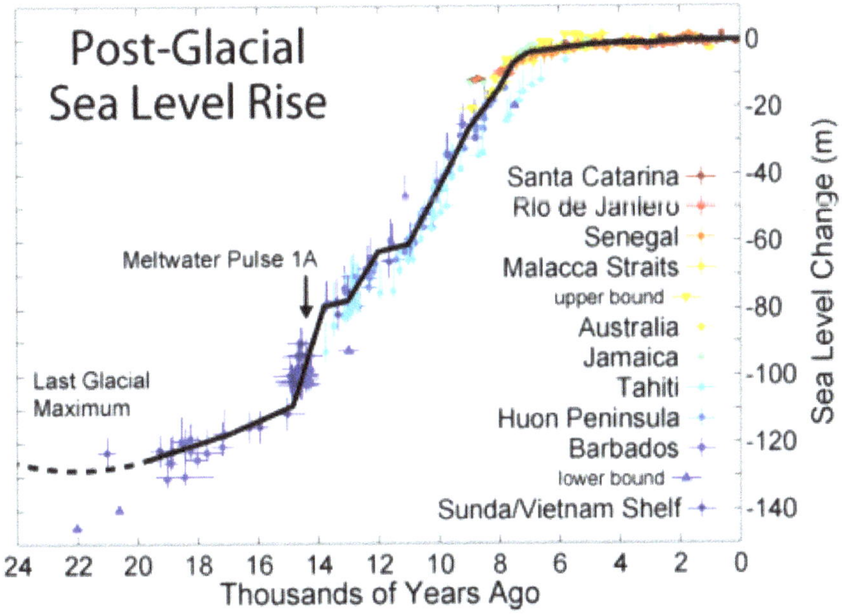

Figure 16.14 Eustatic Sea-Level Curve for the Past 24 ka. Sea-Level rise is Global due to Melting Glaciers. Graph from Wikimedia Foundation Inc., CC BY-SA 3.0.

Isostatic Sea-Level Change

Isostatic sea-level changes are local changes caused by subsidence or uplift of the crust related either to changes in the amount of ice on the land or to growth or erosion of mountains.

Almost all of Canada and parts of the northern United States were covered in thick ice sheets at the peak of the last glaciation. Following the melting of this ice, there has been an isostatic rebound of continental crust in many areas. This ranges from several hundred meters of a rebound in the central part of the Laurentide Ice Sheet (around Hudson Bay) to 100 m to 200 m in the peripheral parts of the

Laurentide and Cordilleran Ice Sheets. In other words, although global sea level was about 130 m lower during the last glaciation, the glaciated regions were depressed at least that much in most places, and more than that in places where the ice was thickest.

Tectonic Sea-Level Change

Tectonic sea-level changes are local changes caused by tectonic processes. The subduction of the Juan de Fuca Plate beneath British Columbia is creating tectonic uplift (about 1 mm/year) along the western edge of Vancouver Island, although much of this uplift is likely to be reversed when the next large subduction-zone earthquake strikes.

Coastlines in areas where there has been a net sea-level rise in the geologically recent past are commonly characterized by estuaries and fiords. A glacially eroded or modified U-shaped valley

that extends below sea level and connects to the ocean. Filled with seawater, depths may reach more than 1,000 feet below sea level. The largest Alaskan fiords are more than 100 miles long and more than 5 miles wide. Also spelled Fiord. The rocky, yet steep coast of Kenai Fjords National Park lies along the southeastern side of the Kenai Peninsula in south-central Alaska.

This region of the Kenai Peninsula is part of the Chugach terrane that collided with Alaska about 65 million years ago; the bedrock is primarily composed of partially metamorphosed muddy sandstones and shales from turbidite deposit with intrusions of granite and granodiorite. The Harding Icefield sits astride the Kenai Mountains and supplies ice to outlet glaciers that carve deep U-shaped valleys into the soft bedrock.

Figure 16.18 Aerial Photo of Puget Sound-- a Fjord. Image by Copernicus Sentinel-2, ESA, CC BY-SA 3.0.

HUMAN INTERFERENCE WITH SHORELINES

There are various modifications that we make in an attempt to influence beach processes for our purposes. Sometimes these changes are effective and may appear to be beneficial, although in most cases there are unintended negative consequences that we don't recognize until much later.

Figure 16.16 The Broad Beach Riprap is Designed to Retain Sand. Image by Ralph Daily, CC BY 2.0.

As shown below, a series of breakwaters (structures parallel to the shore) were built in the 1930s and sand has accumulated behind them to form the bulge on the beach. The breakwaters would have acted as islands and the sand has been deposited in the low-energy water behind them had the breakwater not been worn down.

Figure 16.17 Santa Monica Harbor& Pier, 1936. Image in Introduction to Oceanography by Paul Webb, CC BY 4.0.

An example is at the beach near Malibu, which has been armored with rip-rap and concrete blocks in an attempt to limit the natural erosion that is threatening the properties at the top of the cliff. As already noted, the unintended effect of this installation will be to starve the beach of sediment. As long as the armor remains in place, which might be several decades, there is a risk that the spit will start to erode, which will affect many of the organisms that use that area as their habitat and many of the people who go there for recreation.

Groins have an effect that is similar to that of breakwaters, although groins are constructed perpendicular to the beach, and they trap sediment by slowing the longshore current.

Figure 16.18 New Jersey Groin Fields Interrupt the Flow of Sediment, Worsening Erosion Down the Beach. Image by NOAA is in the public domain.

Most of the sediment that forms beaches along our coasts comes from rivers, so if we want to take care of the beaches, we have to take care of rivers. When a river is dammed, its sediment load is deposited in the resulting reservoir, and for the century or two, while the reservoir is filling up, that sediment cannot get to the sea. During that time, beaches (including spits, baymouth bars, and tombolos) within tens of kilometers of the river's mouth (or more in some cases) are at risk of erosion.

SUMMARY

Waves form when the wind blows over water. The size of the waves depends on the wind speed, the area over which it is blowing, and time. The important parameters of a wave are its amplitude, wavelength, and speed. The water beneath a wave is disturbed to a depth of one- half the wavelength, and a wave is slowed when it approaches shallow water. A longshore current develops where waves approach the shore at an angle, and swash and backwash on a beach move sediment along the shore. The combined effect of these two processes is sediment transport by longshore drift.

Coasts that have experienced uplift within the past several million years tend to have irregular shapes and are dominated by erosional processes. Wave paths are bent where the coast is irregular and wave energy is focused on headlands. Rocky headlands are eroded into sea caves, arches, stacks, and sea cliffs, and the areas around these features are eroded into wave-cut platforms. Over the long term (millions of years), irregular coasts are straightened.

Coasts that have not been uplifted for tens of millions of years tend to be relatively straight, and are dominated by depositional features, although deposition is also important on irregular coasts. Waves and longshore drift are important in controlling the formation of beaches, as well as spits, tombolos, baymouth bars, and barrier islands. Beaches can be divided into zones, such as foreshore and backshore, and beach shapes typically change from season to season.

Carbonate reefs and carbonate sediments form in tropical regions where there is little input of clastic sediments.

The relative levels of the land and sea have significant implications for coastal processes and landforms, and they have been constantly changing over geological time. Eustatic sea-level changes are global in effect and are typically related to glacial ice formation or melting. Isostatic sea-level changes are local effects caused by uplift or subsidence of continental crust, typically because of the gain or loss of glacial ice. Tectonic sea-level changes are related to plate interactions. Net sea-level rise leads to the development of estuaries and fiords, while net sea- level drop creates uplifted marine terraces and beaches.

Humans have a strong urge to alter coasts for their convenience by building seawalls, breakwaters, groynes, and other barriers. Although these types of features may have economic and other benefits, they can have both geological and ecological implications that must be considered.

17 Chapter SHAPED BY RIVERS AND RUNNING WATER

Figure 17.1 The Grand Canyon, Arizona. Image by Jeremy Patrich is used under a CC-BY 4.0 license.

Goals & Objectives of this unit

- Describe a drainage basin and explain the origins of different types of drainage patterns.
- Explain how streams become graded and how certain geological and anthropogenic changes can result in a stream losing their gradation.
- Describe the formation of stream terraces.
- Describe the processes by which sediments are moved by streams and the flow velocities that are necessary to erode them from the stream bed and keep themsuspended in the water.
- Describe the process of stream evolution.

RIVERS

Streams are the most important agents of erosion and transportation of sediments on Earth's surface. They are responsible for the creation of much of the topography that we see around us. They are also locations of great beauty and tranquility, and of course, they provide much of the water that is essential to our existence. But streams are not always peaceful and soothing. During large storms and rapid snowmelts, they can become raging torrents capable of moving cars and houses and destroying roads and bridges. When they spill over their banks, they can flood huge areas, devastating populations, and infrastructure. Over the past century, many of the most damaging natural disasters in California have been attributed to flowing water, and we can expect them to become even more severe as the climate changes.

The Hydrological Cycle

Water is constantly on the move. It is evaporated from the oceans, lakes, streams, the surface of the land, and plants (transpiration) by solar energy. It is moved through the atmosphere by winds and condenses to form clouds of water droplets or ice crystals. It comes back down as rain or snow and then flows through streams, into lakes, and eventually back to the oceans.

Water on the surface and in streams and lakes infiltrates the ground to become groundwater. Groundwater slowly moves through the rock and surficial materials. Some groundwater returns to other streams and lakes, and some go directly back to the oceans.

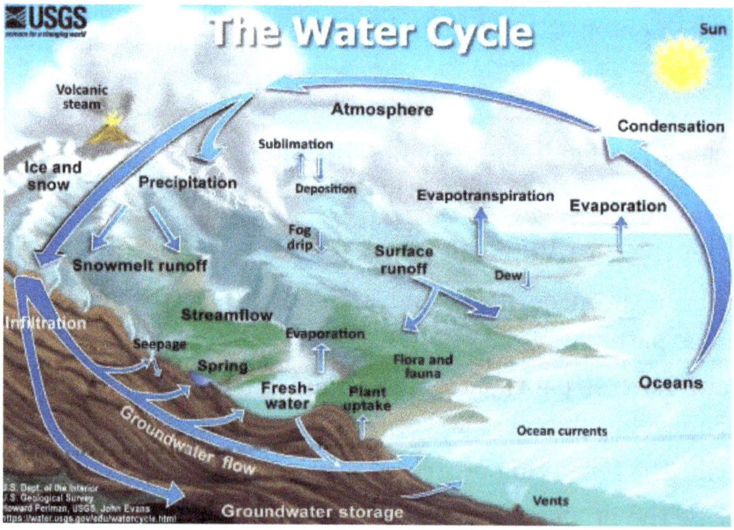

Figure 17.2 The Hydrologic Cycle. Image by USGS is in the public domain.

Even while it's moving around, water is stored in various reservoirs. The largest, by far, is the oceans, accounting for 97% of the volume. Of course, that water is salty. The remaining 3% is freshwater. Two-thirds of our freshwater is stored in ice and one-third is stored in the ground. The remaining freshwater, about 0.03% of the total is stored in lakes, streams, vegetation, and the atmosphere.

Although the proportion of Earth's water that is in the atmosphere is tiny, the actual volume is huge. At any given time, there is the equivalent of approximately 13,000 km3 of water in the air in the form of water vapor and water droplets in clouds. Water is evaporated from the oceans, vegetation, and lakes at a rate of 1,580 km3 per day, and just about the same volume falls as rain and snow every day — over both the oceans and land. The precipitation that falls on land goes back to the ocean in the form of streamflow (117 km3/day) and groundwater flow (6 km3/day).

STREAMS & RIVERS

Freshwater in streams, ponds, and lakes is an extremely important part of the water cycle if only because of its importance to living creatures. Along with wetlands, these freshwater regions contain a tremendous variety of organisms. Streams are bodies of water that have a current; they are in constant motion. Geologists recognize many categories of streams depending on their size, depth, speed, and location. Creeks, brooks, tributaries, bayous, and rivers might all be lumped together as streams. In streams, water always flows downhill, but the form that downhill movement takes varies with rock type, topography, and many other factors. Stream erosion and deposition are extremely important creators and destroyers of landforms and were described in the Erosion and Deposition unit.

Parts of a Stream

A stream originates at its source. A source is likely to be in the high mountains where snows collect in winter and melt in summer, or a source might be a spring. A stream may have more than one source and when two streams come together it's called a confluence.

Figure 17.3 Visual Examples of a Confluence, Tributary, & Estuary. Image by Lumen Learning, CC BY-SA 4.0.

The smaller of the two streams is a tributary of the larger stream. A stream may create a pool where the water slows and becomes deeper. The point at which a stream comes into a large body of water, like an ocean or a lake is called the mouth. Where the stream meets the ocean or lake is an estuary. The mix of fresh and saltwater where a river runs into the ocean creates a diversity of environments where many different types of organisms create unique ecosystems.

Rivers are the largest types of stream, moving large amounts of water from higher to lower elevations. The Amazon River, the world's river with the greatest flow, has a flow rate of nearly 220,000 cubic meters per second! People have used rivers since the beginning of civilization as a source of water, food, transportation, defense, power, recreation, and waste disposal. Also, the lowest elevation a stream can flow is its base level, usually the ocean.

DIVIDES

A divide is a topographically high area that separates a landscape into different water basins. The rain that falls on the north side of a ridge flows into the northern drainage basin and rain that falls on the south side flows into the southern drainage basin. On a much grander scale, entire continent has divides, known as the continental divides.

Figure 17.4: North American Continental Divide (Image on Wikimedia Commons by Pfly, CC BY-SA 3.0)

DRAINAGE BASINS

The drainage pattern is a pattern created by stream erosion over time that reveals characteristics of the kind of rocks and geologic structures in a landscape region drained by streams. The drainage pattern is the pattern formed by the streams, rivers, and lakes in a particular drainage basin. They are governed by the topography of the land, whether a particular region is dominated by hard or soft rocks and the gradient of the land.

A watershed represents all of the stream tributaries that flow to some location along the stream channel. The number, size, and shape of the drainage basins found in an area vary, and the larger the topographic map, the more information on the drainage basin is available. The pattern of tributaries within a drainage basin depends largely on the type of rock beneath, and on structures within that rock (folds, fractures, faults, etc.).

The main types of drainage patterns:

- Dendritic patterns, which are by far the most common, develop in areas where the rock beneath the stream has no particular fabric or structure and can be eroded equally easily in all directions.
- Trellis drainage patterns typically develop where sedimentary rocks have been folded or tilted and then eroded to varying degrees depending on their strength.
- Rectangular patterns develop in areas that have very little topography and a system of bedding planes, fractures, or faults that form a rectangular network.
- Parallel drainage system is a pattern of rivers caused by steep slopes with some relief. Because of the steep slopes, the streams are swift and straight, with very few tributaries, and all flow in the same direction. Parallel drainage patterns form where there is a pronounced slope to the surface. A parallel pattern also develops in regions of parallel, elongate landforms like outcropping resistant rock bands.
- Radial drainage system, the streams radiate outwards from a central high point. Volcanoes usually display excellent radial drainage. Other geological features on which radial drainage commonly develops are domes and laccoliths. On these features, the drainage may exhibit a combination of radial patterns.
- Centripetal drainage system is similar to the radial drainage system, with the only exception that radial drainage flows out versus centripetal drainage flows in.

- Deranged drainage system is a drainage system in drainage basins where there is no coherent pattern to the rivers and lakes.
- Angular drainage patterns form where bedrock joints and faults intersect at more acute angles than rectangular drainage patterns. Angles are both more and less than 90 degrees.

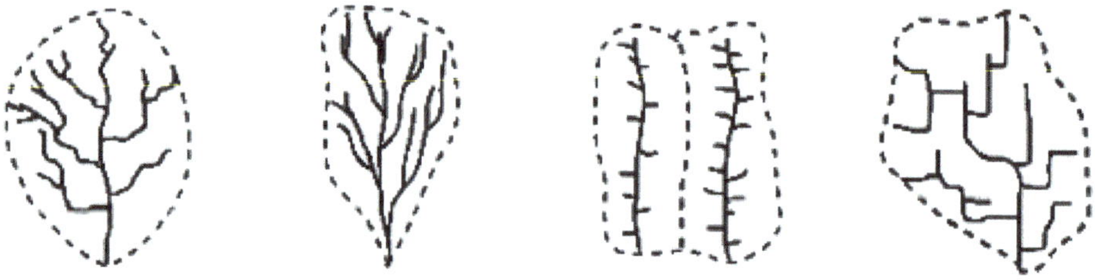

Figure 17.5 Common Drainage Patterns (A) Dendritic, (B) Parallel, (C) Trellis, & (D) Rectangular. Image by Ling Zhang and Eric Guilbert, CC BY 4.0.

RIVER TERRACES

Sediments accumulate within the flood plain of a stream, and then, if the base level changes, or if there is less sediment to deposit, the stream may cut down through those existing sediments to form terraces.

In the late 19th century, American geologist William Davis proposed that streams and the surrounding terrain develop in a cycle of erosion. Following tectonic uplift, streams erode quickly, developing deep V-shaped valleys that tend to follow relatively straight paths.

Gradients are high, and profiles are ungraded. Rapids and waterfalls are common. During the mature stage, streams erode wider valleys and start to deposit thick sediment layers. Gradients are slowly reduced and grading increases. In old age, streams are surrounded by rolling hills, and they occupy wide sediment-filled valleys. Meandering patterns are common.

Figure 17.7 River Terraces along the Colorado River in the Grand Canyon, Arizona. Image by Jeremy Patrich is used under a CC-BY 4.0 license.

Davis's work was done long before the idea of plate tectonics, and he was not familiar with the impacts of glacial erosion on streams and their environments. While some parts of his theory are out of date, it is still a useful way to understand streams and their evolution.

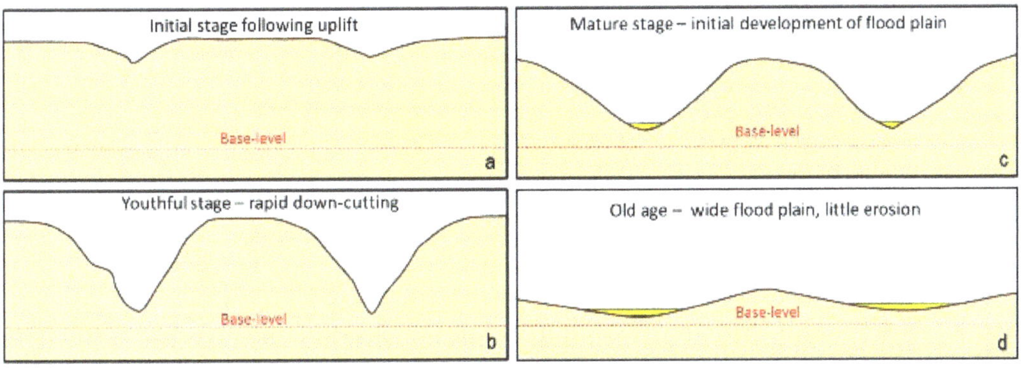

Figure 17.6 A depiction of the Davis cycle of erosion: a: initial stage, b: youthful stage, c: mature stage, and d: old age. Image by Steven Earle (2015) is used under a CC BY 4.0 license.

STREAM EROSION & DEPOSITION

Flowing water is a very important mechanism for both erosion and deposition. Water flow in a stream is primarily related to the stream's gradient, but it is also controlled by the geometry of the stream channel. Water flow velocity is decreased by friction along the stream bed, so it is slowest at the bottom and edges and fastest near the surface and in the middle. The velocity just below the surface is typically a little higher than right at the surface because of friction between the water and the air. On a curved section of a stream, flow is fastest on the outside and slowest on the inside.

A cut bank is the outside bank of a stream, which is continually undergoing erosion. Cut banks are found in abundance along mature or meandering streams, they are located on

the outside of a stream bend. They are shaped much like a small cliff and are formed by the erosion of soil as the stream collides with the riverbank. As opposed to a point bar which is an area of deposition, a cut bank is an area of erosion.

A point bar is a depositional feature made of sediment that accumulates on the inside bend of streams. Point bars are found in abundance in mature or meandering streams. They are crescent-shaped and located on the inside of a stream bend.

Other factors that affect stream-water velocity are the size of sediments on the stream bed — because large particles tend to slow the flow more than small ones — and the discharge, or volume of water passing a point in a unit of time (e.g., m3/second). During a flood, the water level always rises, so there is a greater cross-sectional area for the water to flow in; however, as long as a river remains confined to its channel, the velocity of the water flow also increases.

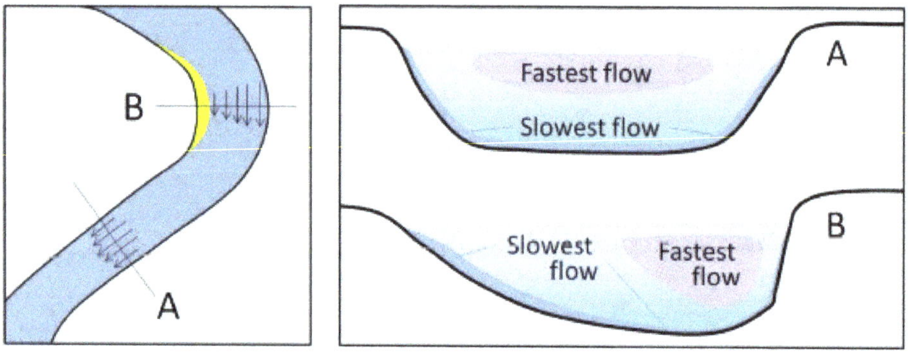

Figure 17.8 The Velocity of a Stream depends on whether the Channel is Straight or Curved. Image by Steven Earle, CC BY 4.0.

Large particles rest on the bottom, bedload, and may only be moved during rapid flows under flood conditions. They can be moved by saltation (bouncing) and by traction (being pushed along by the force of the flow).

Smaller particles may rest on the bottom some of the time, where they can be moved by saltation and traction, but they can also be held in suspension in the flowing water, especially at higher velocities. As you know from intuition and experience, streams that flow fast tend to be turbulent (flow paths are chaotic and the water surface appears rough) and the water may be muddy, while those that flow more slowly tend to have laminar flow (straight-line flow and a smooth water surface) and clear water. Turbulent flow is more effective than laminar flow at keeping sediments in suspension.

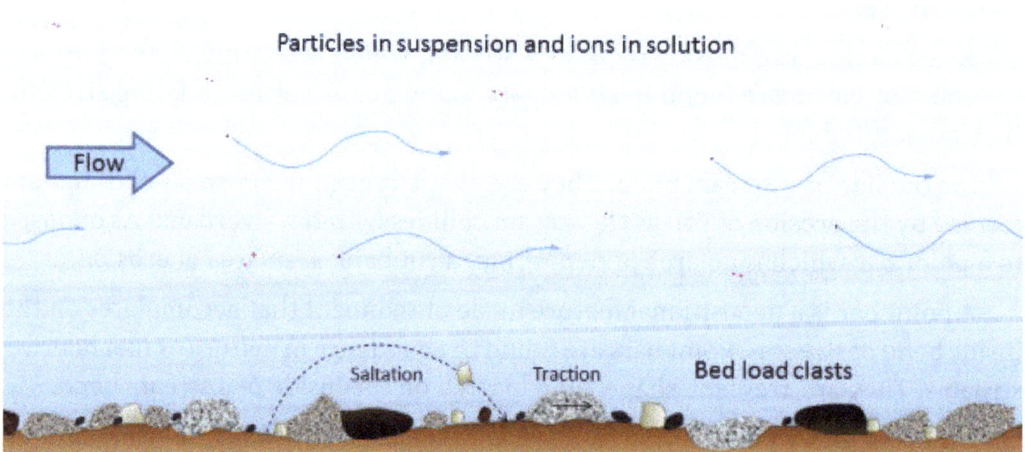

Figure 17.9 Modes of Transportation of Sediment & Dissolved Ions. Image by Steven Earle, CC BY 4.0.

Stream water also has a dissolved load, which represents (on average) about 15% of the mass of material transported and includes ions such as calcium (Ca+2) and chloride (Cl-) in solution.

The faster the water is flowing, the larger the particles that can be kept in suspension and transported within the flowing water. However, as Swedish geographer Filip Hjulström discovered in the 1940s, the relationship between grain size and the likelihood of a grain being eroded, transported, or deposited is not as simple as one might imagine. Consider, for example, a 1 mm grain of sand. If it is resting on the bottom, it will remain there until the velocity is high enough to erode it, around 20 cm/s. But once it is in suspension, that same 1 mm particle will remain in suspension as long as the velocity doesn't drop below 10 cm/s. For a 10 mm gravel grain, the velocity is 105 cm/s to be eroded from the bed but only 80 cm/s to remain in suspension.

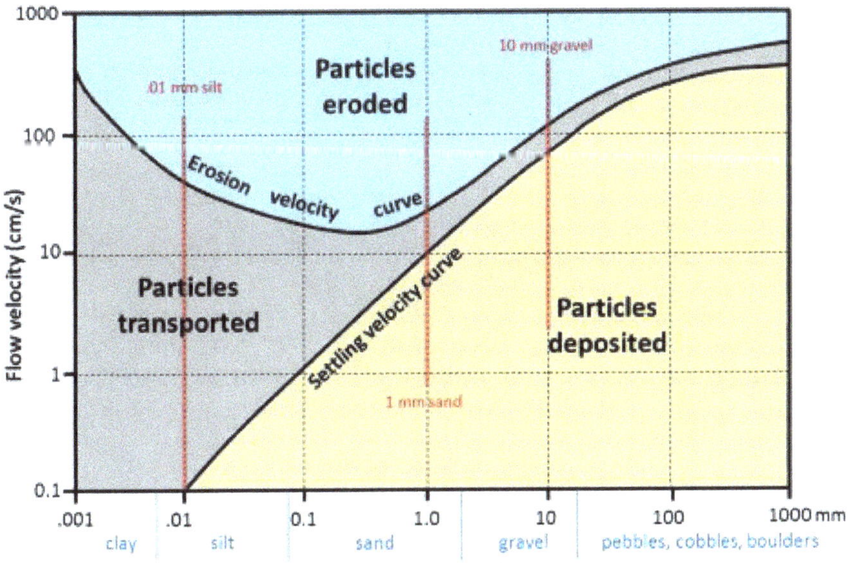

Figure 17.10 A Diagram Showing the Relationships Between Particle Size & Velocity. Image by Steven Earle, CC BY 4.0.

On the other hand, a 0.01 mm silt particle only needs a velocity of 0.1 cm/s to remain in suspension but requires 60 cm/s to be eroded. In other words, a tiny silt grain requires a greater velocity to be eroded than a grain of sand that is 100 times larger! For clay-sized particles, the discrepancy is even greater. In a stream, the most easily eroded particles are small sand grains between 0.2 mm and 0.5 mm. and anything smaller or larger requires a higher water velocity to be eroded and entrained in the flow. The main reason for this is those smaller particles, and especially the tiny grains of clay, have a strong tendency to stick together, and so are difficult to erode from the stream bed.

It is important to be aware that a stream can both erode and deposit sediments at the same time. At 100 cm/s, for example, silt, sand, and medium gravel will be eroded from the stream bed and transported in suspension, coarse gravel will be held in suspension, pebbles will be both transported and deposited, and cobbles and boulders will remain stationary on the stream bed.

A stream typically reaches its greatest velocity when it is close to flooding over its banks. This is known as the bank-full stage. As soon as the flooding stream overtops its banks and occupies the wide area of its flood plain, the water has a much larger area to flow through and the velocity drops significantly. At this point, sediment that was being carried by the high-velocity water is deposited near the edge of the channel, forming a natural bank or levée.

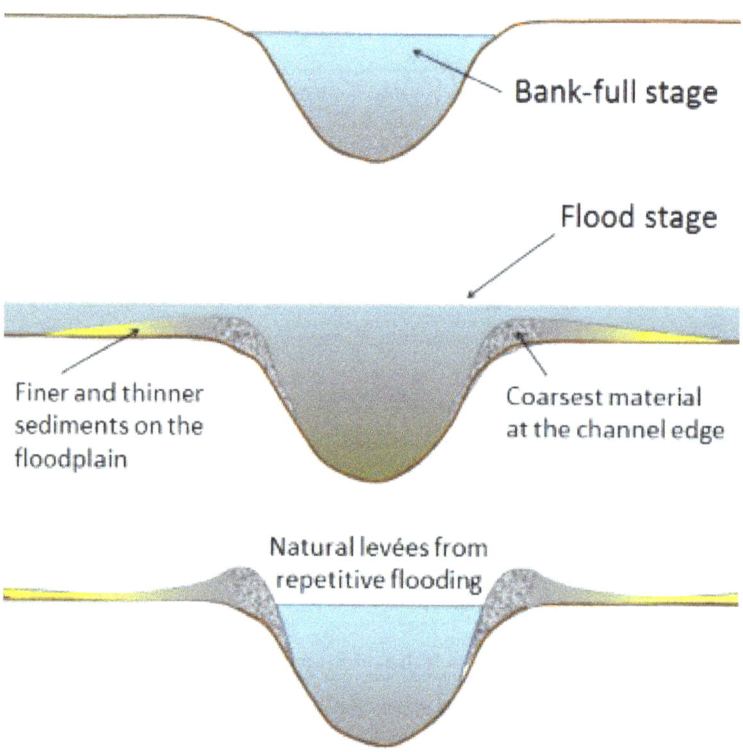

Figure 17.11 The Development of Natural Levées During the Flooding of a Stream. Image by Steven Earle, CC BY 4.0.

Stream Types

Stream channels can be straight or curved, deep and slow, or rapid and choked with coarse sediments. The cycle of erosion has some influence on the nature of a stream, but there are other factors that are equally as important.

Youthful streams that are actively down cutting their channels tend to be relatively straight and are typically ungraded (meaning that rapids and falls are common). Youthful streams commonly have a step-pool morphology, meaning that the stream consists of a series of pools connected by rapids and waterfalls. They also have steep gradients and steep and narrow V-shaped valleys, in some cases steep enough to be called canyons.

Figure 17.12 Waterfall at Whitney Portal, Lone Pine California. Image by Jeremy Patrich is used under a CC-BY 4.0 license.

In mountainous terrains, such as that in the Sierra Nevada, steep youthful streams typically flow into wide and relatively low-gradient U-shaped glaciated valleys. The youthful streams have high sediment loads, and when they flow into the lower-gradient glacial valleys where the velocity isn't high enough to carry all of the sediment, braided patterns develop, characterized by a series of narrow channels separated by gravel bars.

Mature streams water is stored for later use in natural water sources, such as groundwater aquifers, soil water, natural wetlands, and small artificial ponds, tanks, and reservoirs behind major dams. Storing water invites a host of potential issues regardless of that water's intended purpose, including contamination through organic and inorganic means. Mature streams will also introduce a small flood plain, caused by the downcutting and removal of material, and the river channel itself will begin to develop shapes, such as meanders and braiding.

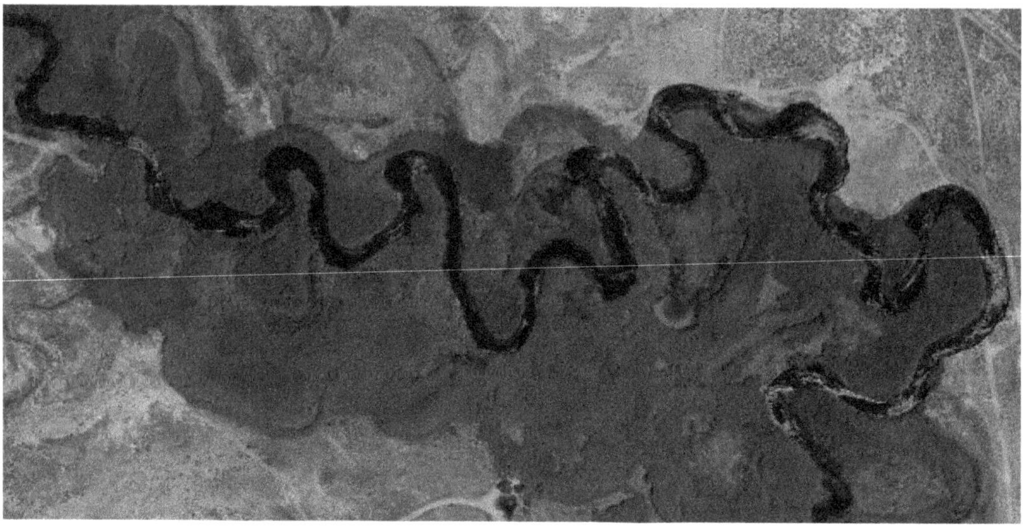

Figure 17.13 The Northern Owens River, The Direction of Flow is Southward, & Note the Abandoned Channels (Map data ©2019 Google).

Braided streams can develop anywhere there is more sediment than a stream can transport. One such environment is in volcanic regions, where explosive eruptions produce large amounts of unconsolidated material that gets washed into streams. The Owens River next to Lone Pine in California is a good example of this.

Old age streams are easily identified by their very low gradient, rich meanders, and a plethora of oxbow lakes and meander scars. The depositional process becomes more predominant and begins forming deltas, estuaries, levees, and distributaries. This landscape will also have a very broad, or wide, floodplain, and low erosive energy. Think of the mighty Mississippi river.

A stream that occupies a wide, flat flood plain with a low gradient typically carries only sand- sized and finer sediments and develops a sinuous flow pattern. When a stream flows around a corner, the water on the outside has farther to go and tends to flow faster. This leads to erosion of the banks on the outside of the curve, deposition on the inside, and formation of a point bar. Over time, the sinuosity of the stream becomes increasingly exaggerated, and the channel migrates around within its flood plain, forming a meandering pattern. The meander in the middle of the photo has reached the point where the thin neck of land between two parts of the channel is about to be eroded through. When this happens, another oxbow lake will form.

WATER STORAGE

Water is stored for later use in natural water sources, such as groundwater aquifers, soil water, natural wetlands, and small artificial ponds, tanks and reservoirs behind major dams. Storing water invites a host of potential issues regardless of that water's intended purpose, including contamination through organic and inorganic means.

Ponds & Lakes

Ponds and lakes are bordered by hills or low rises so that the water is blocked from flowing directly downhill. Ponds are small bodies of freshwater that usually have no outlet; ponds are often fed by underground springs. Lakes are larger bodies of water. Lakes are usually freshwater, although the Great Salt Lake in Utah is just one exception. Water usually drains out of a lake through a river or a stream and all lakes lose water to evaporation.

Figure 17.14: Photo of Students at Lake Sabrina, Bishop California. Image by Jeremy Patrich is used under a CC-BY4.0 license.

Large lakes have tidal systems and currents and can even affect weather patterns. The Great Lakes in the United States contain 22% of the world's fresh surface water. The largest of them, LakeSuperior, has a tide that rises and falls several centimeters each day. The Great Lakes are large enough to alter the weather system in the Northeastern United States by the "lake effect," which is an increase in snow downwind of the relatively warm lakes. The Great Lakes are home to countless species of fish and wildlife. Lakes form in a variety of different ways: in depressions carved by glaciers, in calderas, and along tectonic faults, to name a few. Subglacial lakes are even found below a frozen ice cap.

As a result of geologic history and the arrangement of landmasses, most lakes are in the Northern Hemisphere. More than 60% of all the world's lakes are in Canada—most of these lakes were formed by the glaciers that covered most of Canada in the last Ice Age.

Limnology is the study of bodies of freshwater and the organisms that live there. The ecosystem of a lake is divided into three distinct sections:

- The surface (littoral) zone is the sloped area closest to the edge of the water.
- The open-water zone (also the photic or limnetic zone) has abundant sunlight.
- The deep-water zone (also the aphotic or profundal zone) has little or no sunlight.

There are several life zones found within a lake. In the littoral zone, sunlight promotes plant growth, which provides food and shelter to animals such as snails, insects, and fish. In the open- water zone, other plants and fish, such as bass and trout, live. The deep-water zone does not have photosynthesis since there is no sunlight. Most deep-water organisms are scavengers, such as crabs and catfish that feed on dead organisms that fall to the bottom of the lake. Fungi and bacteria aid in the decomposition in the deep zone.

Though different creatures live in the oceans, ocean waters also have these same divisions based on sunlight with similar types of creatures that live in each of the zones.

Lakes are not permanent features of a landscape. Some come and go with the seasons, as water levels rise and fall. Over a longer time, lakes disappear when they fill with sediments, if the springs or streams that fill them diminish, or if their outlets grow because of erosion. When the climate of an area changes, lakes can either expand or shrink. Lakes may disappear if precipitation significantly diminishes.

Groundwater

Although this may seem surprising, water beneath the ground is commonplace. Usually, groundwater travels slowly and silently beneath the surface, but in some locations, it bubbles to the surface at springs. The products of erosion and deposition by groundwater were described in the Erosion and Deposition chapter. Groundwater is the largest reservoir of liquid freshwater on Earth and is found in aquifers, porous rock, and sediment with water in between. Water is attracted to the soil particles and capillary action, which describes how water moves through a porous media, moves water from wet soil to dry areas.

Aquifers are found at different depths. Some are just below the surface and some are found much deeper below the land surface. A region may have more than one aquifer beneath it and even most deserts are above aquifers. The source region for an aquifer beneath a desert is likely to be far from where the aquifer is located; for example, it may be in a mountain area.

The amount of water that is available to enter groundwater in a region is influenced by the local climate, the slope of the land, the type of rock found at the surface, the vegetation cover, land use in the area, and water retention, which is the amount of water that remains in the ground. More water goes into the ground where there is a lot of rain, flat land, porous rock, exposed soil, and where water is not already filling the soil and rock.

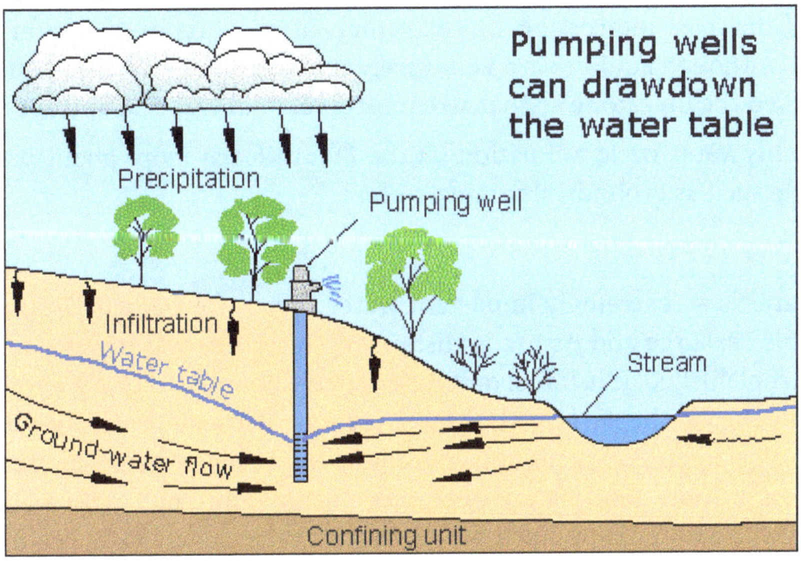

Figure 17.15: Diagram Showing Water Tables, Wells & Ground-Water Flow. Image by USGS is in the public domain.

The residence time of water in a groundwater aquifer can be from minutes to thousands of years. Groundwater is often called "fossil water" because it has remained in the ground for so long, often since the end of the ice ages. To be a good aquifer, it must have good; porosity (small spaces between grains), permeability (connections between pores), and natural recharge by precipitation.

The water droplets are found in the pores between the sediment grains, which is porosity. When the water can travel between ores, that's permeability. To reach an aquifer, surface water infiltrates downward into the ground through tiny spaces or pores in the rock. The water travels down through the permeable rock until it reaches a layer that does not have pores; this rock is impermeable. This impermeable rock layer forms the base of the aquifer. The upper surface where the groundwater reaches is the water table.

The Water Table

For a groundwater aquifer to contain the same amount of water, the amount of recharge must equal the amount of discharge. What are the likely sources of recharge? What are the likely sources of discharge? In wet regions, streams are fed by groundwater; the surface of the stream is the top of the water table. In dry regions, water seeps down from the stream into the aquifer. These streams are often dry much of the year. Water leaves a groundwater reservoir in streams or springs. People take water from aquifers, too. What happens to the water table when there is a lot of rainfall? What happens when there is a drought? Although groundwater levels do not rise and fall as rapidly as at the surface, over time the water table will rise during wet periods and fall during droughts.

One of the most interesting, but extremely atypical types of aquifers is found in Florida. Although aquifers are very rarely underground rivers, in Florida water has dissolved the limestone so that streams travel underground and aboveground.

View this water cycle animation by the EPA to learn more about parts of the water cycle such as groundwater and aquifers.

GROUNDWATER USE

Groundwater is an extremely important water source for people. Groundwater is a renewable resource and its use is sustainable when the water pumped from the aquifer is replenished. It is important for anyone who intends to dig a well to know how deep beneath the surface the water table is. Because groundwater involves interaction between the Earth and the water, the study of groundwater is called hydrogeology. Some aquifers are overused; people pump out more water than is replaced. As the water is pumped out, the water table slowly falls, requiring wells to be dug deeper, which takes more money and energy. Wells may go completely dry if they are not deep enough to reach into the lowered water table. The Ogallala Aquifer supplies about one-third of the irrigation water in the United States. The aquifer is found from 30 to 100 meters deep over about 440,000 square kilometers! The water in the aquifer is mostly from the last ice age. The Ogallala Aquifer is widely used by people for municipal and agriculturalneeds. About eight times more water is taken from the Ogallala Aquifer each year than is replenished. Much of the water is used for irrigation of crops in the Bread Basket of the central plains. Currently, there is great concern about the long-term health of this vast aquifer because it is being tapped into and used at a greater rate than being replenished by natural processes.

This could have huge implications in regard to food production in the country if this critical water source is depleted. At current rates of use, 70% of the aquifer could be gone by 2050. Overuse and lowering of the water tables of aquifers could have other impacts as well.

Lowering the water table may cause the ground surface to sink. Subsidence may occur beneath houses and other structures. When coastal aquifers are overused, saltwater from the ocean may enter the aquifer, contaminating the aquifer and making it less useful for drinking and irrigation. Saltwater incursion is a problem in developed coastal regions, such as in Hawaii.

Springs & Wells

Groundwater meets the surface in a stream, as shown below, or a spring. A spring may be constant or may only flow at certain times of the year. Towns in many locations depend on water from springs. Springs can be an extremely important source of water in locations where surface water is scarce.

A well is created by digging or drilling beneath the surface, to reach groundwater. When the water table is close to the surface, wells are a convenient method for extracting water. When the water table is far below the surface, specialized equipment must be used to dig a well. Most wells use motorized pumps to bring water to the surface, but some still require people to use a bucket to draw water up.

Finally, it's also important to understand how water is cleaned, filtered, and delivered to our homes and work. Too many of us do not know where our water comes from and we take it for granted. This often leads to the wasteful water use of our lawns, showers, and other appliances.

View this website for more rules and regulations to learn more about parts of the water treatment and the EPA.

Figure 17.16 An Image Showing Both a Natural Spring & Man-Made Well. Image by Lumenlearning, CC BY-SA 4.0.

SUMMARY

Water is stored in the oceans, glacial ice, the ground, lakes, rivers, and the atmosphere. Its movement is powered by the sun and gravity.

All of the precipitation that falls within a drainage basin flows into the stream that drains that area. Stream drainage patterns are determined by the type of rock within the basin. Over geological time, streams change the landscape that they flow within, and eventually, they become graded, meaning their profile is a smooth curve. A stream can lose that gradation if there is renewed uplift or if their base level changes for some reason.

Youthful streams in steep areas erode rapidly, and they tend to have steep, rocky, and relatively straight channels. Where sediment-rich streams empty into areas with lower gradients, braided streams can form. In areas with even lower gradients, and where silt and sand are the dominant sediments, meanders are common. Deltas form where streams flow into standing water.

18 Chapter

SHAPED BY WIND AS A GEOMORPHIC AGENT

Figure 18.1 Mesquite Dunes, Death Valley California. Image by Jeremy Patrich is used under a CC-BY 4.0 license.

Goals & Objectives of this unit

- To identify and explain desert landforms, and whether they are formed by erosion and deposition.
- Describe the main types of dunes and the conditions that form them.
- Understand how wind erosion changes and defines landscapes.
- Understand how sand supply and wind variability can design dunes and dune systems.
- Describe the factors that control and explain the processes of wind erosion, transportation, and deposition.

AEOLIAN PROCESSES

Aeolian processes are most effective where surface material is fine, dry, and loose. Vegetation imposes a frictional force on the wind to reduce its effectiveness. Thus, an environment devoid of vegetation is best. These conditions are met in deserts found on every continent of the world. In most cases, wind erosion predominates over deposition leaving a surface of stones. Only one-quarter of Earth's deserts are partially or completely covered by sand.

Though less extensive in area, coastlines of large bodies of water are another aeolian environment. Here, waves and currents supply weathered material susceptible to wind action. Aeolian processes have been enhanced by human activity over the past few centuries, especially in Asia and Africa. Overuse of soil and grazing land resources in semi-arid and arid and seasonally dry regions has led to extensive wind erosion and desertification.

EROSIONAL FEATURES

The entire solid surface of the Earth is subject to wind erosion. It is the balance between the driving force of the wind and the resistance of the surface that ultimately determines whether surface materials are detached and transported away.

The roughness length of the surface, a parameter based on the size and distance between objects in a group, has a critical control over the velocity of the wind. Grass, shrubs, and trees, all impart a drag on the wind to reduce its erosive force. Vegetation also acts to bind soil particles to the surface. A surface without a cover of vegetation exposes soil to the direct force of the wind making erosion more effective. Thus, dry regions lacking a protective cover of vegetation display the effects of wind erosion more than humid climates.

The cohesiveness of surface materials affects the resistance of the surface to erosion. Clay particles exhibit a greater degree of cohesiveness than sand. Clay-rich soils tend to resist erosion by wind more than less cohesive materials. Thus, clays require a much higher threshold velocity for detachment than one would expect. Cohesiveness may be provided by a cementing agent as well. Calcium carbonate and other salts common to desert regions deposited in interparticle voids harden and bind particles together.

Features of Wind Erosion

Landscapes formed from the work of wind result from either the removal of fine particles or the sculpting effects of material in movement. Deflation lifts and removes loose particles from the surface. Deserts, where soils of mixed particle size have been eroded of fines, leave a cobblestone-like surface behind called desert lag or desert pavement. The interlocking pavement of stones protects the underlying surface from the wind. If disturbed, the surface becomes subject to erosion. Such has been the result of surface degradation from military activities in the desert regions of North Africa during World War II (as well as later conflicts fought in the desert regions of the Middle East).

Deflation may also result in blowout depressions, basins ranging in size from less than a meter to many meters deep. Chemical weathering of cementing materials loosens a particles cohesiveness to enable wind erosion. Some extremely large depressions like the Qattara Depression in the western desert of Egypt are partially a result of deflation.

Sand particles lifted free of the surface can sandblast rock surfaces in a process called abrasion. Abrasion shapes and polishes exposed bedrock. Abrasion is restricted to a distance of about a meter or two above the surface because sand grains are lifted a short distance. Ventifacts are smooth faceted rocks that often have a polished surface that results from abrasion. At a much larger scale, elongate ridges called yardangs form by the abrasion and streamlining rock structures oriented parallel to the prevailing wind direction. Abrasion occurs at the windward end while deflation removes material from the leeward end. Archaeologists have suggested that the Great Sphinx of Giza, Egypt is partially formed from a yardang.

Sand Transportation

Much of our knowledge about sand transport dates back to the 1930s and 1940s when British Brigadier Ralph A. Bagnold conducted experiments while stationed in the desert of North Africa. Sand generally begins to move when the wind achieves a velocity of about 4.5 m/sec. At first, sand exhibits a rolling motion called traction or surface creep. Approximately 20 to 25% of total sand transport during sand storms occurs by traction. As wind speed increases, grains are lifted into the air by wind gusts.

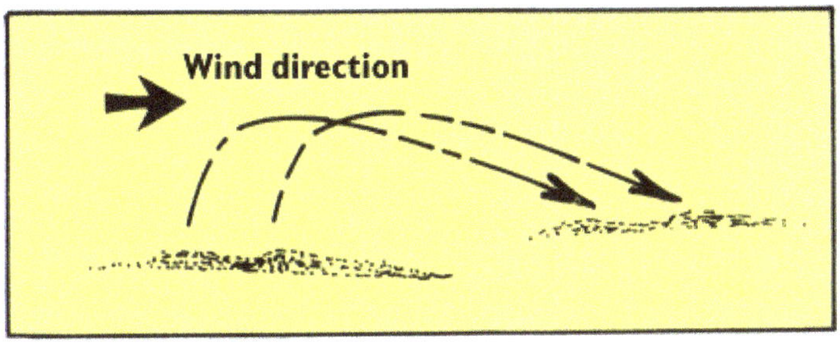

Figure 18.2 Example of Saltating Sand Grains. Image by USGS is in the public domain.

Once airborne, sand grains travel downwind and then drop back to the surface several centimeters from their point of origin. Finer dust particles are lifted from the surface and suspended in the air at much greater heights than heavier sand grains. With strong winds and turbulence, sand grains can be lifted as high as two meters and travel a distance of 10 meters or more. When a settling sand grain impacts the surface, it sends another grain of sand into the air to travel in the downwind direction. Watching this process in action makes sand appear to be bouncing along the surface. Saltation accounts for over 50% of sand transport over dunes. Once saltation begins, transport can continue under somewhat lower wind speeds.

Dust storms are primarily composed of fine material that easily reduces visibilities to a few meters and can persist for hours. During a prolonged dry period, extraordinary amounts of valuable topsoil were stripped from the Great Plains during the "Dust Bowl" period of the 1930s. Moister conditions returned, and the application of soil conservation techniques have greatly reduced the erosive effects of wind. Sandstorms are raging systems of blowing, stinging sand traveling a meter or two above the surface.

A strange phenomenon that has mystified generations is the sound moving sand can make as it is transported across the surface. In approximately 30 places around the world, a sustained hum rings out from this resonating dune when wind conditions are right. Geoscientists have long wondered what causes booming dunes.

DEPOSITIONAL FEATURES & DUNE TYPES

A sand dune is an accumulation of loose sand grains piled up by the wind. Dunes are most likely to form where winds are strong and generally blow from the same direction. Some of the most extensive dune fields are found in the world's great deserts like the Sahara. The dune fields of these great sandy deserts are called sand seas. Patches of dunes are found in the southwest desert of the United States. Ancient dune fields are found in regions that were dry in the past but now exhibit a more humid climate. Dune systems are found landward of beaches where sand is blown landward. Beach dunes are common along the coasts of the world's oceans and large lakes like the Great Lakes of North America.

Dune Formation

Dunes are primarily made up of accumulated sand grains. Finer silt and clay are carried further by the wind leaving the heavier sand grains behind. Generally, a dune forms an asymmetric cross-sectional form with a gentle windward or stoss slope and a steep leeward slope called the slip face. Saltating grains move up the windward slope and then come to rest at their angle of repose on the downwind side in a zone of stagnant air.

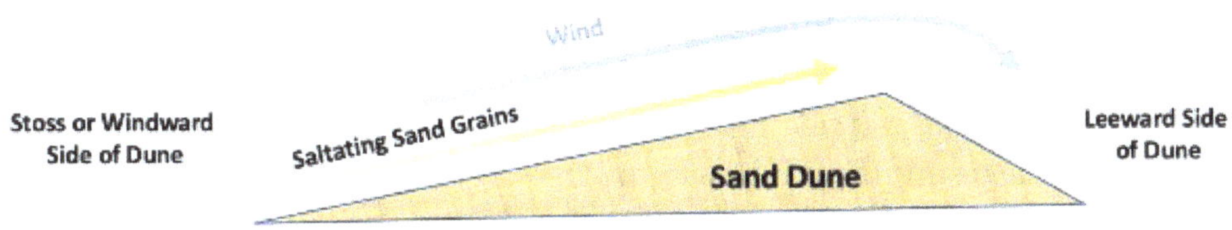

Figure 18.3 Sand Movement Along a Dune. Figure by Jeremy Patrich is licensed under CC BY-NC-SA 4.0

SAND

While deserts are defined by dryness, not sand, the popular conception of a typical desert is a sand sea called an erg. An erg is a broad area of desert covered by a sheet of fine-grained sand often blown by aeolian forces (wind) into dunes. Probably the best-known erg is the Empty Quarter (Rub' al Khali) of Saudi Arabia, but other ergs exist in Colorado (Great

Sand Dunes National Park), Utah (Little Sahara Recreation Area), New Mexico (White Sands National Monument), and California (parts of Death Valley National Park). It is not only deserts that form dunes; the high supply of sand can form ergs anywhere, even as far north as 60° in Saskatchewan at the Athabasca Sand Dunes Provincial Park. Coastal ergs on the shores of lakes and oceans also do exist and can be found in places like Oregon, Michigan, and Indiana.

The way dunes form creates an internal feature called cross-bedding. As the wind blows up the windward side of the dune, it carries sand to the dune crest depositing layers of sand parallel to the windward (or "stoss") side. The sand builds up the crest of the dune and pours over the top until the leeward (downwind or slip) face of the dune reaches the angle of repose, the maximum angle which will support the sand pile. Dunes are unstable features and move as the sand erodes from the stoss side and continues to drop down the leeward side covering previous stoss and slip-face layers and creating the cross-beds. Mostly, these are reworked over and over again, but occasionally, the features are preserved in a depression, then lithified. Shifting wind directions and abundant sand sources create chaotic patterns of cross-beds like those seen in the fossil ergs represented by the Navajo Sandstone and Zion National Park of Utah.

In the Mesozoic, Utah was covered by a series of ergs, thickest in Southern Utah. Perhaps the best known of these sandstone formations is the Navajo Sandstone. The Navajo forms the dramatic cliffs and spires in Zion National Park and covers a large part of the Colorado Plateau. It is exposed beneath the Entrada Sandstone in Arches National Park, a later series of dunes in which the conditions of the lithified rock allowed the formation of arches.

As the cement that hold the grains together in these modern sand cliffs disintegrate and the freed grains gather at the base of the cliffs and move down the washes, sand grains may be recycled and redeposited. These great sand ergs may represent ancient quartz sands recycled many times, just passing now through another cycle. One example of this is Coral Pink

Sand Dunes State Park in Southwestern Utah, which is sand that is eroded from the Navajo Sandstone forming new dunes.

Figure 18.4 Enlarged Image of Sand Grains from Coral Pink Sand Dunes, Utah. Image by Utah State Parks Office is in the public domain.

Dunes

Sand dunes are formed by wind moving sand particles. The shape of the dune varies with the amount of sand available and the direction that the wind blows. If the wind blows steadily from one direction, linear, transverse, or barchan dunes will form. If the direction that the wind blows shifts, star or network dunes will form. The diagram below notes the different types of dunes possible based on sand supply and wind variability.

A barchan or crescent dune forms as individual units commonly found in dry regions lacking vegetation. The elongated tips, or horns, point in the downwind direction. Barchans are found moving across a non-sandy surface of gravel or clay up of several tens of meters a year.Symmetrical barchan dunes indicate winds blowing from a constant direction.

Transverse dunes form perpendicular to the prevailing wind as accumulations of loose, well- sorted, very fine to medium sand. They have a gentle stoss slope (usually less than 15°) and a steep (32°) slip face.

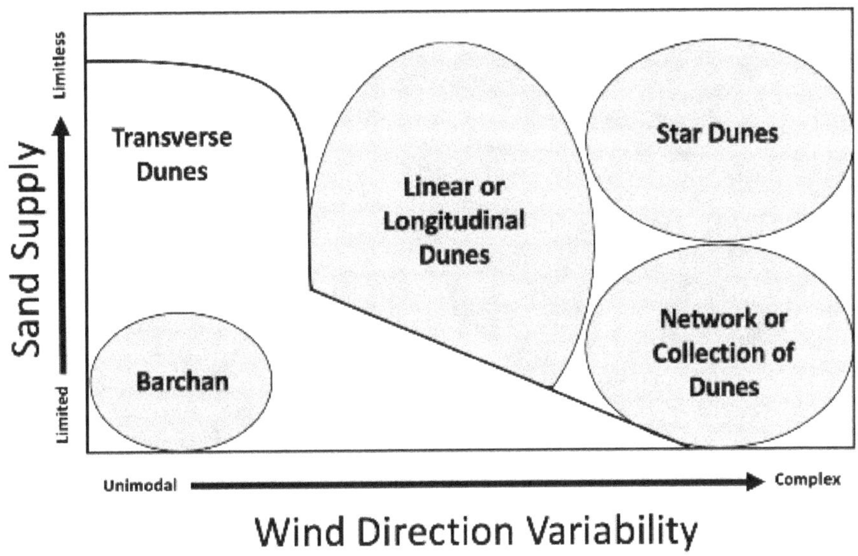

Figure 18.5 Image Separating Sand Dunes by Wind Variability and Sand Supply. Figure by Jeremy Patrich is licensed under CC BY-NC-SA 4.0

Individual dunes whose horns point in the upwind direction are parabolic dunes. These hairpin- like dunes form where sand accumulates in a moist environment with a cover of vegetation. Vegetation or dampness in the lower part slows the dune motion there, permitting the dry crest to push ahead of its base causing the horns to be anchored behind.

Blowout dunes form from sand deposited at the end of an open-ended deflation hollow. Blowout dunes are common to many sandy coasts (coastal blowout dunes). The deflation hollow focuses the wind toward the center of the depression causing the sand to migrate landward as a dune. With little wind energy on the side of the hollow, the tails of the dune move much slower. As a result, beach plants like dune grass, sea oats or sand cherry can find asuitable habitat to flourish in. Once established, the plants help anchor the flanks to promote the formation of the characteristic parabolic shapeLongitudinal dunes form where the wind blows from more than one direction in a region with an abundant supply of sand. In some places, longitudinal dunes form long smooth whalebacks, while in other regions they display knife-edged crests. Unlike transverse dunes, these do not migrate but lengthen in size over time. Seifs are elongate dunes that resemble great windrows of sand. They form parallel to the direction of the wind, though are a product of shifting wind direction. The changing wind direction sends sand back and forth across the dune crest leaving a sharp knife-like profile.

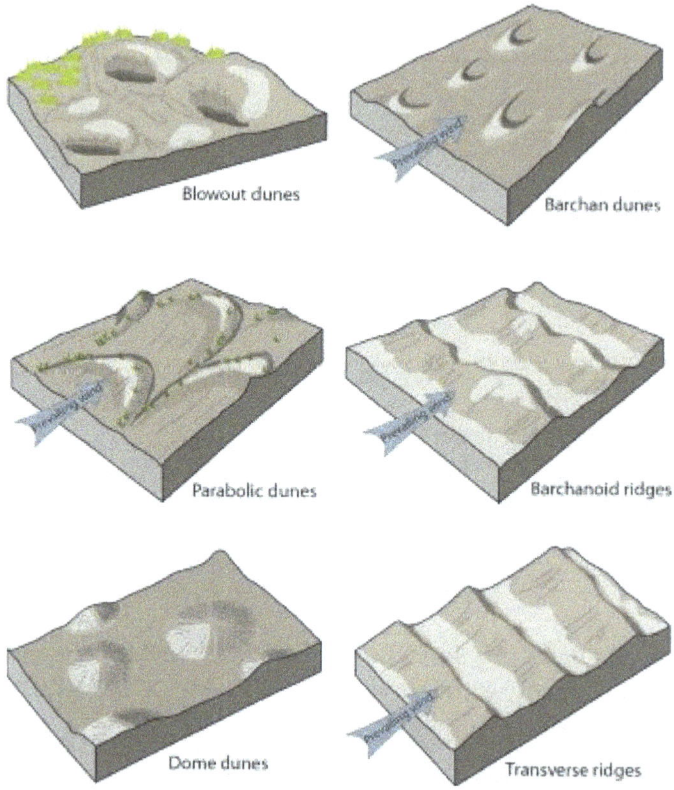

Figure 18.6 Five Common Dunes, Note Their Wind Direction. Image by Trista L. Thornberry-Ehrlich, Colorado State University after Fryberger, S.G., L.F. Krystinik, and C.J. Schenk. 1990; and McKee, E.D. 1966, is in the public domain.

ERGS

Individual sand seas are referred to as ergs. Small ergs are present in most desert regions. North Africa and the Arabian Desert contain the most spectacular forms on Earth with waves of sand covering tens of thousands of square kilometers. Ergs are also found in India, western China, and Australia. Star dunes are pyramidal-shaped mounds of sand with slipfaces on three or more arms radiating from the center and common to ergs. They tend to accumulate in areas where the wind comes from several different directions. Star dunes grow vertically rather than laterally. The Sand Hills of Nebraska are remnants of an ancient sand sea.

ALLUVIAL FANS

Rivers are important features in any ecosystem. Apart from providing water for the different needs, it can also be dragged by gravity across the Earth to form some of the most beautiful landmarks, especially in the valleys. The flowing waters carry with it the alluvium across its path. The alluvium consists of soil, and small rock particles picked up by the river as it flows. The alluvium is deposited on flat plains where smaller streams fan out across the plain. Over time the deposits stand out from the surrounding landscape to form a unique feature called alluvial fans. The alluvial fans formed range from just a minuscule to a massive sculpture that can be seen from the space. Alluvial fans are either cone or fan-shaped in nature. When an alluvial fan is built by debris the flow, then it is referred to as the debris cone or colluvial fan.

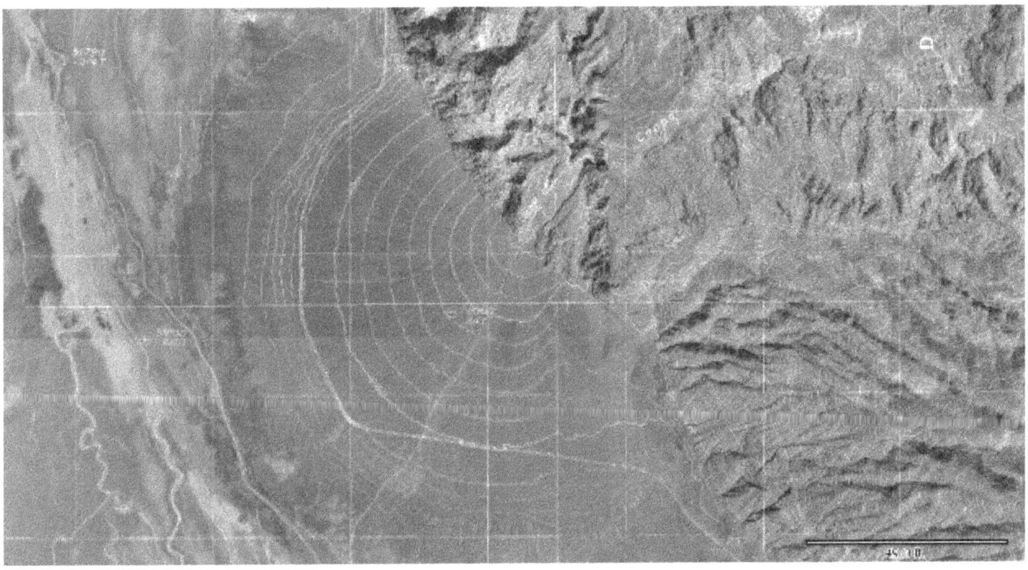

Figure 18.7 Alluvial Fan in Death Valley. Image on Wikimedia Commons is in the Public Domain.

Alluvial fans are mainly formed by rivers but can also be formed by streams. The rivers and streams gather sediments or particles of soil and small rocks as it flows towards a particular direction. As the gradient of the river or stream decreases, it drops small rock particles reducingthe capacity of the channel which forces the river to change course. The stream or river gradually builds a slightly mounded conical fan shape with the sediments deposited at the apex of the fan. Alluvial fans are mostly coarse-grained, especially at their mouth but are relatively fine-grained at their edges.

Alluvial fans are common in desert areas characterized by flash floods in the nearby hills. The watercourses in desert areas are large funnel-shaped basin at the top which opens up to the alluvial fan. Alluvial fans are also common in a wet climate. The Koshi River in Nepal has built a mega alluvial fan which covers an area of 15,000 square kilometers at the point where the river traverses into India. The streams flowing into California Central Valley in North America have also deposited some sediment forming alluvial fans. The largest alluvial fan in the world is formed in the Taklimakan desert in China. The alluvial fan is 56.6 km wide and 61.3 km long with parts of the fan alive with water which flows from the Molcha River. Apart from the Earth's surface, alluvial fans are also found on Mars and other parts of the solar system.

BAJADAS

Alluvial fans formed on plains may also coalesce along a mountain to form a feature that is commonly referred to as Bajada. Bajada is a Spanish word that is often used to describe a landscape or geomorphology and mean inclination or descent. Bajada is common in a dry climate such as the Southwestern US where the flash floods deposit alluvium over time to form a series of alluvial fans that coalesce to form the Bajada. Bajadas can also be formed on wetter climate with streams continuously depositing sediments.

Figure 18.8 A Collection of Alluvial Fans are Called Bajadas, Death Valley National Park, California. Used under Google Earth reproduction guidelines

OTHER GEOMORPHIC ARID FEATURES

Rain does fall occasionally in deserts, and desert storms are often violent. A record 44 millimeters of rain once fell within 3 hours in the Sahara. Large Saharan storms may deliver up to 1 millimeter per minute. Normally dry stream channels, called arroyos or wadis, can quickly fill after heavy rains, and flash floods make these channels dangerous. More people drown in deserts than die of thirst.

Though little rain falls in deserts, deserts receive runoff from ephemeral, or short-lived, streams fed by rain and snow from adjacent highlands. These streams fill the channel with a slurry of mud and commonly transport considerable quantities of sediment for a day or two. Although most deserts are in basins with closed, or interior drainage, a few deserts are crossed by 'exotic' rivers that derive their water from outside the desert. Such rivers infiltrate soils and evaporate large amounts of water on their journeys through the deserts, but their volumes are such that they maintain their continuity. The Nile, The Colorado, and The Yellow Rivers are exotic rivers that flow through deserts to deliver their sediments to the sea.

Lakes form where rainfall or meltwater in interior drainage basins is sufficient. Desert lakes are generally shallow, temporary, and salty. Because these lakes are shallow and have a low bottom gradient, wind stress may cause the lake waters to move over many square kilometers. When small lakes dry up, they leave a salt crust or hardpan. The flat area of clay, silt, or sand encrusted with salt that forms is known as a playa. There are more than a hundred playas in North American deserts. Most are relics of large lakes that existed during the last Ice Age about 12,000 years ago. Lake Bonneville was a 52,000-square-kilometer lake almost 300 meters deep in Utah, Nevada, and Idaho during the Ice Age. Today the remnants of Lake Bonneville include Utah's Great Salt Lake, Utah Lake, and Sevier Lake. Because playas are arid landforms from a wetter past, they contain useful clues to climatic change.

Buttes

In geomorphology, a butte is an isolated hill with steep, often vertical sides and a small, relatively flat top; buttes are smaller landforms than mesas, plateaus, and tablelands. The word "butte" comes from a French word meaning "small hill"; its use is prevalent in the Western United States, including the southwest where "mesa" is used for the larger landform. Because of their distinctive shapes, buttes are frequently landmarks in plains and mountainous areas. In differentiating mesas and buttes, geographers use the rule of thumb that a mesa has a top that is wider than its height, while a butte has a top that is narrower than its height.

Figure 18.9 Devils Tower, Wyoming. Image by Bradley Davis, CC BY 2.0.

Mesas

Mesa (Spanish, Portuguese and Sardinian for table) is the American English term for tableland, an elevated area of land with a flat top and sides that are usually steep cliffs. It takes its name from its characteristic table-top shape. It may also be called a table hill, table-topped hill or, Table Mountain. It is larger than a butte, which it otherwise resembles closely.

It is a characteristic landform of arid environments, particularly the Western and Southwestern United States in badlands and mountainous regions ranging from Washington and California to the Dakotas, Wyoming, Utah, Oklahoma, and Texas. Examples are also found in many other nations including Spain, Sardinia, North and South Africa, Arabia, India, and Australia.

Mesas form by weathering and erosion of horizontally layered rocks that have been uplifted by tectonic activity. Variations in the ability of different types of rock to resist weathering and erosion cause the weaker types of rocks to be eroded away, leaving the more resistant types of rocks topographically higher than their surroundings. This process is called differential erosion.

The most resistant rock types include sandstone, conglomerate, quartzite, basalt, chert, limestone, lava flows, and sills. Lava flows and sills, in particular, are very resistant to weathering and erosion, and often form the flat top, or caprock, of a mesa. The less resistant rock layers are mainly made up of shale, a softer rock that weathers and erodes more easily.

Figure 18.10 Aerial View of Mesas in Monument Valley, Arizona. Image on Wikipedia by brewbooks, CC BY-SA 2.0.

The differences in the strength and resilience of various rock layers are what give mesas their distinctive shape. Less resistant rocks are eroded away on the surface into valleys, where they collect water drainage from the surrounding area, while the more resistant layers are left standing out. A large area of very resistant rock, such as a sill may shield the layers below it from erosion while the softer rock surrounding it is eroded into valleys, thus forming a caprock.

Differences in rock type also reflect on the sides of a mesa, as instead of smooth slopes, the sides are broken into a staircase pattern called cliff-and-bench topography. The more resistant layers form the cliffs, or stairsteps, while the less resistant layers form gentle slopes, or benches, between the cliffs. Cliffs retreat and are eventually cut off from the main cliff, or plateau, by basal sapping. When the cliff edge does not retreat uniformly, but instead is indented by headward eroding streams, a section can be cut off from the main cliff, forming a mesa.

Basal sapping occurs as water flowing around the rock layers of the mesa erodes the underlying soft shale layers, either as surface runoff from the mesa top or from groundwater moving through permeable overlying layers, which leads to slumping and flowage of the shale. As the underlying shale erodes away, it can no longer support the overlying cliff layers, which collapse and retreat. When the caprock has caved away to the point where only little remains, it is known as a butte.

Figure 18.11 The Pajarito Volcanic Plateau in New Mexico. Image by Patricksfisher1, CC BY-SA 4.0

Plateaus

In geology and physical geography, a plateau is also called a high plain or a tableland, is an area of highland, usually consisting of relatively flat terrain, which is raised significantly above the surrounding area, often with one or more sides with steep slopes. Plateaus can be formed by several processes, including upwelling of volcanic magma, extrusion of lava, and erosion by water and glaciers. Plateaus are classified according to their surrounding environment as intermontane, piedmont, or continental.

SUMMARY

The arid environment can be defined as one in which the amount of precipitation an area receives, divided by the amount, which is lost to evapotranspiration. The arid environment can be organized into three zones: hyper-arid, arid, and semi-arid. The zones are characterized by low annual precipitation.

Dunes are made of sand-sized particles, and may consist of quartz, calcium carbonate, as sand is just a descriptor of the grain size. The upwind side of the dune is called the stoss side; the downflow side is called the lee side. Sand is pushed (creep) or bounces (saltation) up the stoss side and slides down the lee side. A side of a dune that the sand has slid down is called a slip face (or slipface).

The differences between buttes, mesas and plateaus are their surficial size.

19 Chapter

SHAPED BY GLACIERS

Figure 19.1 Field Trip at Lake Sabrina, Bishop California. Image by Jeremy Patrich is used under a CC-BY 4.0 license.

Goals & Objectives of this unit

- Describe the timing and extent of Earth's past glaciations, going as far back as the earlyProterozoic.
- Describe the important geological events that led up to the Pleistocene glaciations and how the Milankovitch orbital variations controlled the timing of those glaciations.
- Explain the differences between continental and alpine glaciation.
- Describe and identify the various landforms related to alpine glacial erosion, including U-shaped valleys, arêtes, cols, horns, hanging valleys, truncated spurs, and drumlins.

- Identify various types of glacial lakes, including tarns, Finger Lakes, and kettle lakes.

GLACIERS

A glacier is a long-lasting body of ice (decades or more) that is large enough (at least tens of meters thick and at least hundreds of meters in extent) to move under its own weight. About 10% of Earth's land surface is currently covered with glacial ice, and although the vast majority of that is in Antarctica and Greenland, there are many glaciers in Canada, especially in the mountainous parts of Alberta, and Yukon and in the far north. At various times during the past million years, glacial ice has been much more extensive, covering at least 30% of the land surface at times.

Glaciers represent the largest repository of freshwater on Earth (~69% of all freshwater), and they are highly sensitive to changes in climate. In the current warming climate, glaciers are melting rapidly worldwide, and although some of the larger glacial masses will last for centuries more, many smaller glaciers will be gone within decades, and in some cases, within years.

Glaciers have long played a role in the geologic history of California. In the past, glaciers were active in several areas of California, leaving behind geologic evidence such as glacial deposits, mountain cirques, and glacial striations. In northern California, evidence from glaciers during the Pleistocene is found at Mount Shasta, Lassen Volcano, throughout the Klamath Mountains, Medicine Lake Volcano, and the Coast ranges. In central California evidence from glaciers can be found in the Sierra Nevada and The White Mountains. The only evidence of glaciation in southern California has been found in the San Bernardino Mountains.

During the cold Pleistocene epoch (Ice Age), which lasted between 1.8 million years ago to 10,000 years ago, glaciers fluctuated in size. Geologic evidence in the Sierra Nevada suggests at least five major glacial periods: McGee, Sherwin, Tahoe, Tioga, and Recess Peak. Similar evidence of glacier fluctuation is found in other regions of California. In the Holocene, the climate began to cool around 1350 A.D., during which time glaciers began to grow. This cool period, referred to as the Little Ice Age, persisted roughly through 1850 A.D., when glaciers are thought to have reached their maximum extent. Since that time, mountain glaciers in California and throughout most of the world have shown signs of overall recession during the past century.

Glacial Periods in Earth's History

We are currently in the middle of a glacial period (although it's less intense now than it was 20,000 years ago) but this is not the only period of glaciation in Earth's history; there have been many in the distant past. In general, however, Earth has been warm enough to be ice-free for much more of the time than it has been cold enough to be glaciated.

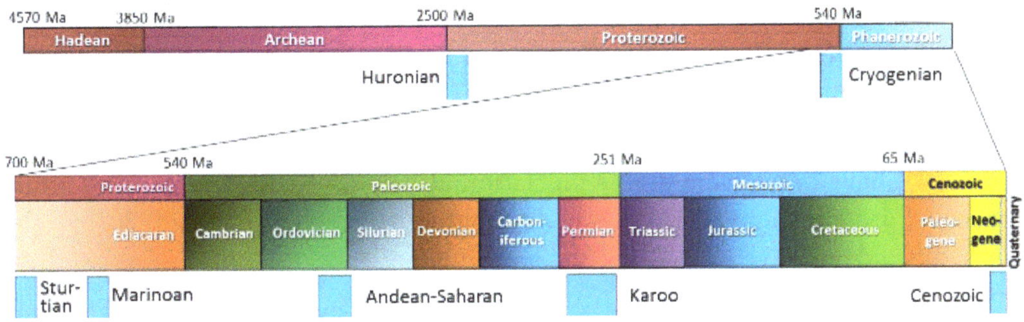

Figure 19.2 The Record of Major Past Glaciations during Earth's History. Image by Steven Earle, CC BY 4.0.

The oldest known glacial period is the Huronian. Based on the evidence of glacial deposits from the area around Lake Huron in Ontario and Michigan, it is evident that the Huronian Glaciation lasted from approximately 2,400 to 2,100 Ma. Because rocks of that age are rare, we don't know much about the intensity or the global extent of this glaciation.

Late in the Proterozoic, for reasons that are not fully understood, the climate cooled dramatically, and Earth was seized by what appears to be its most intense glaciation. The glaciations of the Cryogenian Period (cryo is Latin for icy cold) are also known as the "Snowball Earth" glaciations because it is hypothesized that the entire planet was frozen — even in equatorial regions — with ice on the oceans up to 1 km thick. A visitor to our planet at that time might not have held out much hope for its inhabitability, although life still survived in the oceans. There were two main glacial periods within the Cryogenian, each lasting for about 20 million years: the Sturtian at around 700 Ma and the Marinoan at 650 Ma. There is also evidence of some shorter glaciations both before and after these. The end of the Cryogenian glaciations coincides with the evolution of relatively large and complex life forms on Earth. This started during the Ediacaran Period, and then continued with the so-called explosion of life forms in the Cambrian. Some geologists think that the changing environmental conditions of the Cryogenian are what triggered the evolution of large and complex life.

There have been three major glaciations during the Phanerozoic (the past 540 million years), including the Andean/Saharan (recorded in rocks of South America and Africa), the Karoo (named for rocks in southern Africa), and the Cenozoic glaciations. The Karoo was the longest of the Phanerozoic glaciations, persisting for much of the time that the supercontinent Gondwana was situated over the South Pole (~360 to 260 Ma). It covered large parts of Africa, South America, Australia, and Antarctica. As you might recall from Chapter 10, this widespread glaciation, across continents that are now far apart, was an important component of Alfred Wegener's evidence for continental drift. Unlike the Cryogenian glaciations, the Andean/Saharan, Karoo, and Cenozoic glaciations only affected parts of Earth. During Karoo times, for example, what is now North America was near the equator and remained unglaciated.

Earth was warm and essentially unglaciated throughout the Mesozoic. Although there may have been some alpine glaciation at this time, there is no longer any record of it. The dinosaurs, which dominated terrestrial habitats during the Mesozoic, did not have to endure icy conditions.

A warm climate persisted into the Cenozoic; there is evidence that the Paleocene (~50 to 60 Ma) was the warmest part of the Phanerozoic since the Cambrian. Several tectonic events during the Cenozoic contributed to persistent and significant planetary cooling since 50 Ma. For example, the collision of India with Asia and the formation of the Himalayan range and the Tibetan Plateau resulted in a dramatic increase in the rate of weathering and erosion. Higher than normal rates of weathering of rocks with silicate minerals, especially feldspar, consumes carbon dioxide from the atmosphere and therefore reduces the greenhouse effect, resulting in long-term cooling.

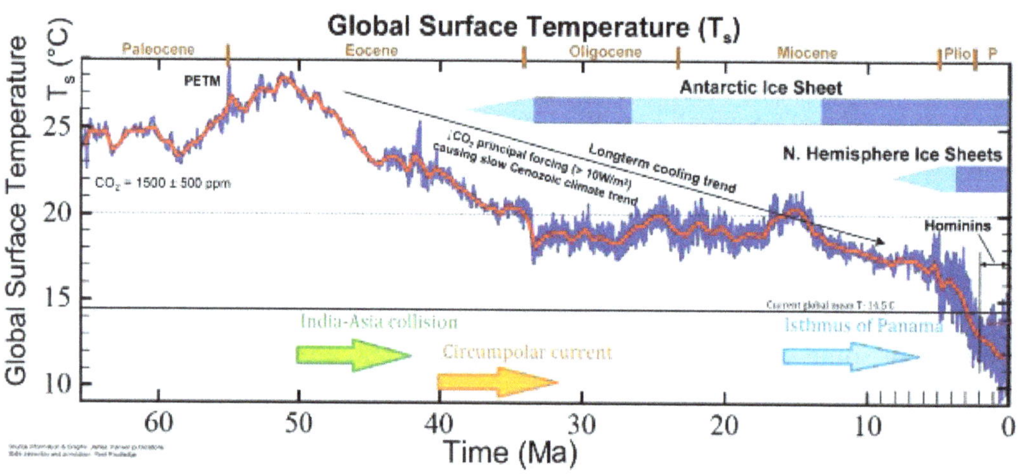

Figure 19.3 The Global Temperature Trend over the Past 65 Ma, the Cenozoic. Image by Steven Earle, CC BY 4.0.

At 40 Ma, ongoing plate motion widened the narrow gap between South America and Antarctica, resulting in the opening of the Drake Passage. This allowed for the unrestricted west-to-east flow of water around Antarctica, the Antarctic Circumpolar Current, which effectively isolated the Southern Ocean from the warmer waters of the Pacific, Atlantic, and Indian Oceans. The region cooled significantly, and by 35 Ma (Oligocene) glaciers had started to form on Antarctica.

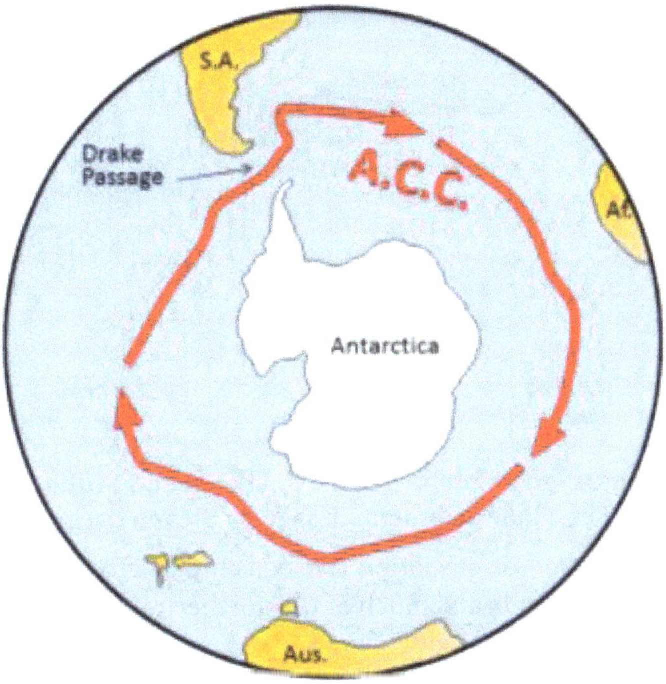

Figure 19.4 The Antarctic Circumpolar Current Prevents Warm Water from the Rest of Earth's Oceans from Traveling to Antarctica. Image by Steven Earle, CC BY 4.0.

Global temperatures remained relatively steady during the Oligocene and early Miocene, and the Antarctic glaciation waned during that time. At around 15 Ma, subduction-related volcanism between central and South America created the connection between North and South America, preventing water from flowing between the Pacific and Atlantic Oceans. This further restricted the transfer of heat from the tropics to the poles, leading to a rejuvenation of the Antarctic glaciation. The expansion of that ice sheet increased Earth's reflectivity enough to promote a positive feedback loop of further cooling: more reflective glacial ice, more cooling, more ice, etc. By the Pliocene (~5 Ma) ice sheets had started to grow in North America and northern Europe. The most intense part of the current glaciation — and the coldest climate — has been during the past million years (the last one-third of the Pleistocene), but if we count Antarctic glaciation, it really extends from the Oligocene to the Holocene, and will likely continue.

The Pleistocene has been characterized by significant temperature variations (through a range of almost 10°C) on time scales of 40,000 to 100,000 years, and corresponding expansion and contraction of ice sheets. These variations are attributed to subtle changes in Earth's orbital parameters (Milankovitch cycles), which are explained in more detail in this unit. Over the past million years, the glaciation cycles have been approximately 100,000 years.

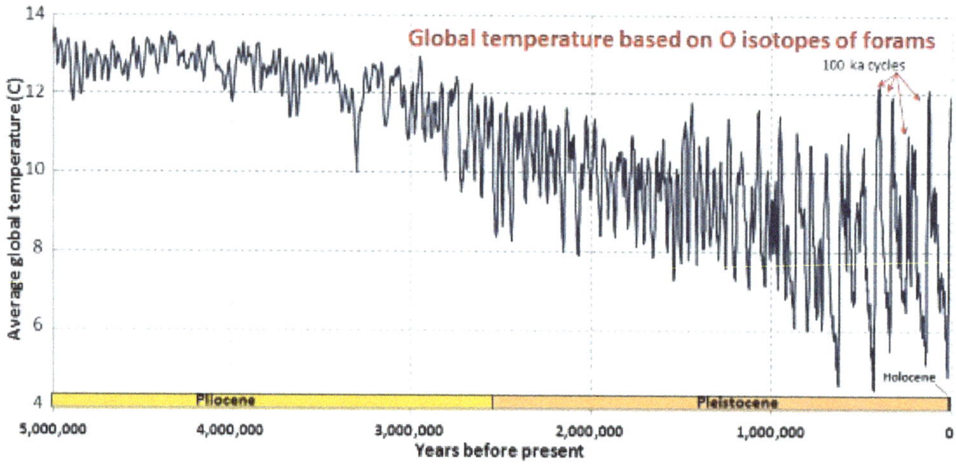

Figure 19.5 Foram Oxygen Isotope Record for the past 5 Million Years, Based on Data from Sea-Floor Sediments.Image by Steven Earle, CC BY 4.0.

At the height of the last glaciation (Wisconsin Glaciation), massive ice sheets covered almost all of Canada and much of the northern United States. The massive Laurentide Ice Sheet covered most of eastern Canada, as far west as the Rockies, and the smaller Cordilleran Ice Sheet covered most of the western region. At various other glacial peaks during the Pleistocene and Pliocene, the ice extent was similar to this, and in some cases, even more extensive. The combined Laurentide and Cordilleran Ice Sheets were comparable in volume to the current Antarctic Ice Sheet.

Figure 19.6 The Extent of the Cordilleran &Laurentide Ice Sheets during the Wisconsin Glaciation. Image by Steven Earle, CC BY 4.0.

HOW GLACIERS WORK

There are two main types of glaciers. Continental glaciers cover vast areas of land in extreme Polar Regions, including Antarctica and Greenland. Alpine glaciers, otherwise known as valley glaciers, originate on mountains, mostly in temperate and Polar Regions, but even in tropical regions if the mountains are high enough.

Earth's two great continental glaciers, on Antarctica and Greenland, comprise about 99% of all of the world's glacial ice, and approximately 68% of all of Earth's freshwater. The Antarctic Ice Sheet is vastly bigger than the Greenland Ice Sheet; it contains about 17 times as much ice. If the entire Antarctic Ice Sheet were to melt, sea level would rise by about 80 m and most of Earth's major cities would be submerged.

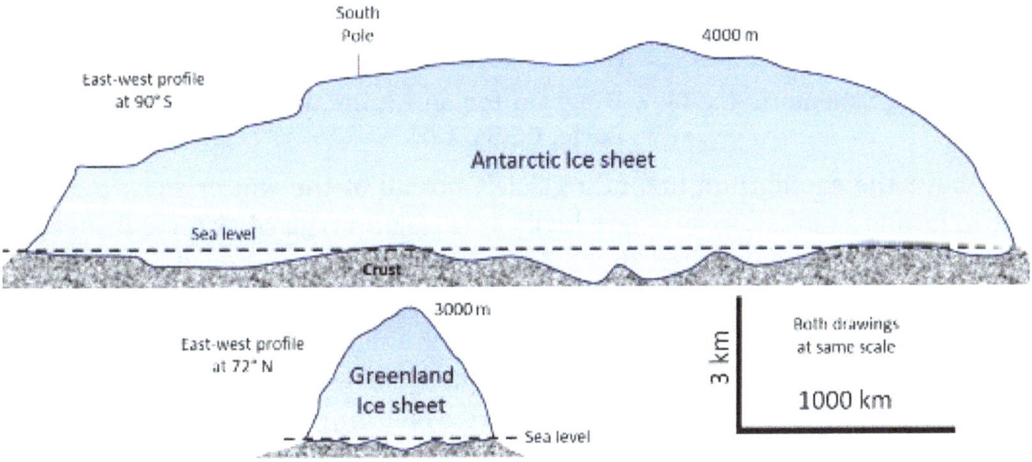

Figure 19.7 Simplified Cross-Sectional Profiles of the Continental Ice Sheets in Greenland & Antarctica. Image by Steven Earle, CC BY 4.0.

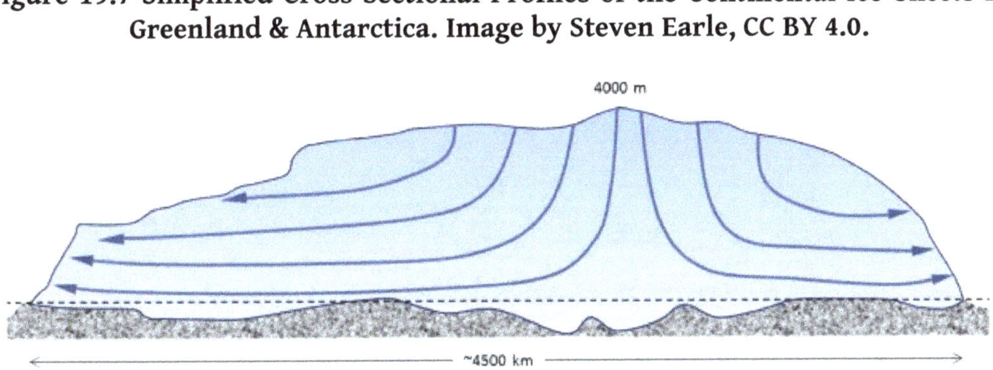

Figure 19.8 Schematic Ice-Flow Diagram. Image by Steven Earle, CC BY 4.0.

The flow of alpine glaciers is primarily controlled by the slope of the land beneath the ice. In the zone of accumulation, the rate of snowfall is greater than the rate of melting. In other words, not all of the snow that falls each winter melts during the following summer, and the ice surface is always covered with snow. In the zone of ablation, more ice melts than accumulates as snow. The equilibrium line marks the boundary between the zones of accumulation (above) and ablation (below).

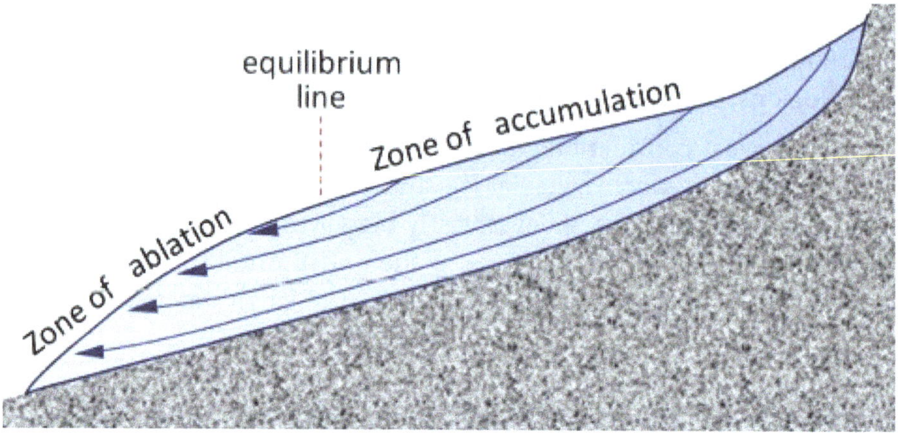

Figure 19.9 Schematic Ice-Flow Diagram for an Alpine Glacier. Image by Steven Earle, CC BY 4.0.

Above the equilibrium line of a glacier, not all of the winter snow melts in the following summer, so snow gradually accumulates. The snow layer from each year is covered and compacted by subsequent snow, and it is gradually compressed and turned into firn within which the snowflakes lose their delicate shapes and become granules. With more compression, the granules are pushed together, and the air is squeezed out. Eventually, the granules are "welded" together to create glacial ice. Downward percolation of water from melting taking place at the surface contributes to the process of ice formation.

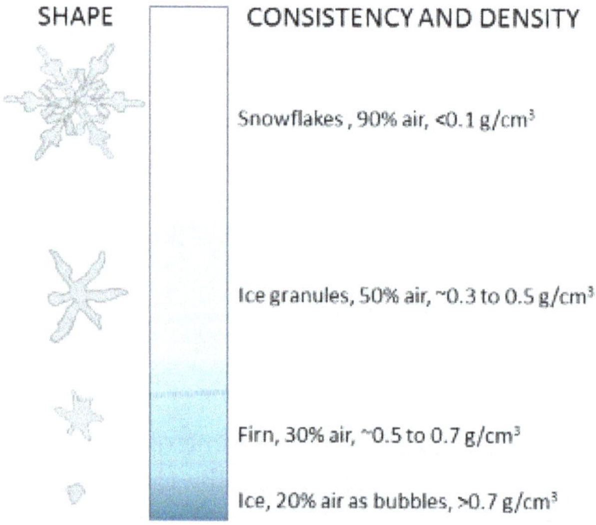

Figure 19.10 Steps in the Process of Formation of Glacial Ice. Image by Steven Earle, CC BY 4.0.

The equilibrium line of a glacier near Whistler, British Columbia., is seen below. Below that line, in the zone of ablation, bare ice is exposed because last winter's snow has all melted; above that line, the ice is still mostly covered with snow from last winter. The position of the equilibrium line changes from year to year as a function of the balance between snow accumulation in the winter and snowmelt during the summer. More winter snow and less summer melting favor the advance of the equilibrium line (and of the glacier's leading edge), but of these two variables, it is the summer melt that matters most to a glacier's budget. Cool summers promote glacial advance and warm summers promote glacial retreat.

Figure 19.11 The Equilibrium Line in 2013 on the Overlord Glacier. Image by Steven Earle, CC BY 4.0.

Glaciers move because the surface of the ice is sloped. This generates stress on the ice, which is proportional to the slope and to the depth below the surface. The stresses are quite small near the ice surface but much larger at depth, and also greater in areas where the ice surface is relatively steep. Ice will deform, meaning that it will behave in a plastic manner, at stress levels of around 100 kilopascals; therefore, in the upper 50 m to 100 m of the ice, flow is not plastic (the ice is rigid), while below that depth, ice is plastic and will flow.

When the lower ice of a glacier flows, it moves the upper ice along with it, so although it might seem from the stress patterns (red numbers and red arrows) that the lower part moves the most, in fact while the lower part deforms (and flows) and the upper part doesn't deform at all, the upper part moves the fastest because it is pushed along by the lower ice.

The plastic lower ice of a glacier can flow like a very viscous fluid and can therefore flow over irregularities in the base of the ice and around corners. However, the upper rigid ice cannot flow in this way, and because it is being carried along by the lower ice, it tends to crack where the lower ice has to flex. This leads to the development of crevasses in areas where the rate of flow of the plastic ice is changing. In the area shown below, for example, the glacier is speedingup over the steep terrain, and the rigid surface ice has to crack to account for the change in velocity.

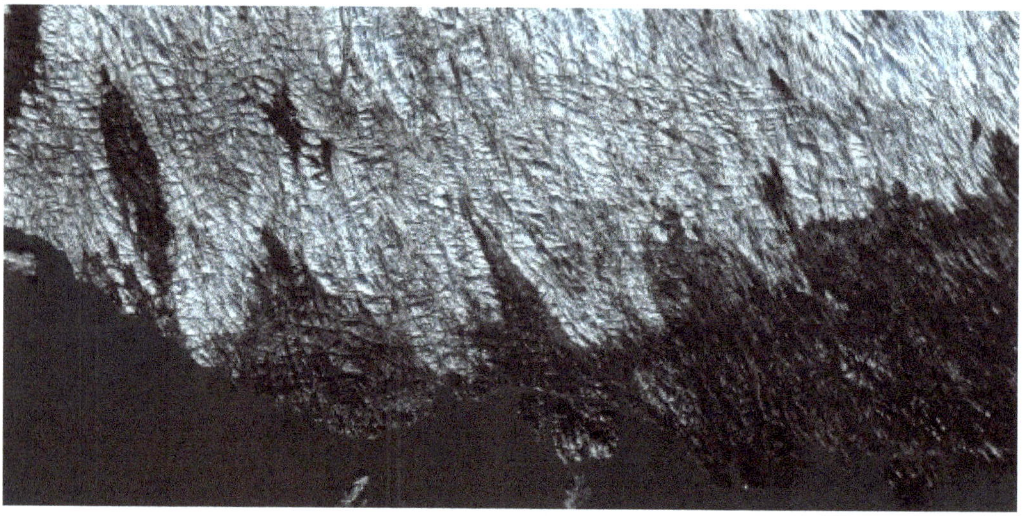

Figure 19.12 Crevasses on a Glacier in Alaska. used under Google Earth reproduction guidelines

The base of a glacier can be cold (below the freezing point of water) or warm (above the freezing point). If it is warm, there will likely be a film of water between the ice and the material underneath, and the ice will be able to slide over that surface. This is known as basal sliding. If the base is cold, the ice will be frozen to the material underneath and it will be stuck — unable to slide along its base. In this case, all of the movement of the ice will be by the internal flow.

One of the factors that affect the temperature at the base of a glacier is the thickness of the ice. Ice is a good insulator. The slow transfer of heat from Earth's interior provides enough heat to warm up the base of the ice is thick, but not enough if it is thin and that heat can escape. It is typical for the leading edge of an alpine glacier to be relatively thin, so it is common for that part to be frozen to its base while the rest of the glacier is still sliding. Because the leading edge of the glacier is stuck to its frozen base, while the rest continues to slide, the ice coming from behind has pushed (or thrust) itself over top of the part that is stuck fast.

Glacial ice always moves downhill, in response to gravity, but the front edge of a glacier is always either melting or calving into water (shedding icebergs).

If the rate of forwarding motion of the glacier is faster than the rate of ablation (melting), the leading edge of the glacier advances (moves forward). If the rate of forwarding motion is about the same as the rate of ablation, the leading edge remains stationary, and if the rate of forwarding motion is slower than the rate of ablation, the leading-edge retreats (moves backward).

Glacial Erosion

Glaciers are effective agents of erosion, especially in situations where the ice is not frozen to its base and can therefore slide over the bedrock or other sediment. The ice itself is not particularly effective at erosion because it is relatively soft (Mohs hardness 1.5 at 0°C); instead, it is the rock fragments embedded in the ice and pushed down onto the underlying surfaces that do most of the erosion. A useful analogy would be to compare the effect of a piece of paper being rubbed against a wooden surface, as opposed to a piece of sandpaper that has embedded angular fragments of garnet.

The results of glacial erosion are different in areas with continental glaciation versus alpine glaciation. Continental glaciation tends to produce relatively flat bedrock surfaces, especially where the rock beneath is uniform in strength. In areas where there are differences in the strength of rocks, a glacier tends to erode the softer and weaker rock more effectively than the harder and stronger rock. Much of central and eastern Canada, which was completely covered by the huge Laurentide Ice Sheet at various times during the Pleistocene, has been eroded to a relatively flat surface. In many cases, the existing relief is due to the presence of glacial deposits such as drumlins, eskers, and moraines (all discussed below) rather than to differential erosion.

ALPINE GLACIERS

Alpine glaciers produce very different topography than continental glaciers, and much of the topographic variability of western Canada can be attributed to glacial erosion. In general, glaciers are much wider than rivers of similar length, and since they tend to erode more at their bases than their sides, they produce wide valleys with relatively flat bottoms and steep sides known as U-shaped valleys. Yosemite National Park was occupied by a large glacier. Glacial systems reached depths of up to 4,000 feet (1,200 m) and left their marks in the Yosemite area. The longest glacier in the Yosemite area ran down the Grand Canyon of the Tuolumne River for 60 miles (97 km), passing well beyond HetchHetchy Valley. Merced Glacier flowed out of Yosemite Valley and into the Merced River Gorge. Lee Vining Glacier carved Lee Vining Canyon and emptied into Lake Russel (the much-enlarged ice age version of Mono Lake). Only the highest peaks, such as Mount Dana and Mount Conness, were not covered by glaciers.

Retreating glaciers often left recessional moraines that impounded lakes such as the 5.5 miles (9 km) long Lake Yosemite

Figure 19.13 Areal View of Yosemite Valley. used under Google Earth reproduction guidelines

U-shaped valleys and their tributaries provide the basis for a wide range of alpine glacial topographic features, examples of which are visible on the International Space Station view of the Swiss Alps. This area was much more intensely glaciated during the last glacial maximum. At that time, the large U-shaped valley in the lower right was occupied by glacial ice, and all of the other glaciers shown here were longer and much thicker than they are now. But even at the peak of the Pleistocene Glaciation, some of the higher peaks and ridges would have been exposed and not directly affected by glacial erosion. A peak that extends above the surrounding glacier is called a nunatak. In these areas, and the areas above the glaciers today, most of the erosion is related to freeze-thaw effects.

Figure 19.14 A View from the International Space Station of the Swiss Alps. Image by Steven Earle, CC BY 4.0.

Some of the important features visible in the image below are arêtes: sharp ridges between U- shaped glacial valleys; cols: low points along arêtes that constitute passes between glacial valleys; horns: steep peaks that have been glacially and freeze-thaw eroded on three or more sides; cirques: bowl-shaped basins that form at the head of a glacial valley; hanging valleys: U- shaped valleys of tributary glaciers that hang above the main valley because the larger main- valley glacier eroded more deeply into the terrain; and truncated spurs (a.k.a. "spurs"): the ends of arêtes that have been eroded into steep triangle-shaped cliffs by the glacier in the corresponding main valley.

Figure 19.15 Bower Island, a Drumlin in Howe Sound, Canada. Image by Steven Earle, CC BY 4.0.

Several other glacial erosion features exist at smaller scales. For example, a drumlin is an elongated feature that is streamlined at the down-ice end. The image below shows a drumlin, and is larger than most, and is made up almost entirely of rock. Drumlins made up of glacial sediments are very common in some areas of continental glaciation

Figure 19.16 Glacial Striations at Mount Rainier National Park. Image by Walter Siegmund, CC BY-SA 3.0.

A rochemoutonée is another type of elongated erosional feature that has a steep and sometimes jagged down-ice end. On a smaller scale still, glacial grooves (tens of centimeters to meters wide) and glacial striations (millimeters to centimeters wide) are created by fragmentsof rock embedded in the ice at the base of a glacier. Glacial striations are very common on rock surfaces eroded by both alpine and continental glaciers.

Glacial Lakes

Lakes are common features in glacial environments. A lake that is confined to a glacial cirque is known as a tarn. Tarns are common in areas of alpine glaciation because the ice that forms a cirque typically carves out a depression in bedrock that then fills with water. In some cases, a series of such basins will form, and the resulting lakes are called rock basin lakes or paternoster lakes.

Figure 19.17 Note the Cirque the Envelops Lake Sabrina, a Glacial Tarn. Image by Jeremy Patrich is used under a CC-BY 4.0 license.

A lake that occupies a glacial valley, but is not confined to a cirque, is known as a finger lake. In some cases, a finger lake is confined by a dam formed by an end moraine, in which case it may be called a moraine lake.

In areas of continental glaciation, the crust is depressed by the weight of glacial ice that is up to 4,000 m thick. Basins are formed along the edges of continental glaciers (except for those that cover entire continents like Antarctica and Greenland), and these basins fill with glacial meltwater. Many such lakes, some of them huge, existed at various times along the southern edge of the Laurentide Ice Sheet. One example is Glacial Lake Missoula, which formed within Idaho and Montana, just south of the B.C. border with the United States. During the latter part of the last glaciation (30 ka to 15 ka), the ice holding back Lake Missoula retreated enough to allow some of the lake water to start flowing out, which escalated into a massive and rapid outflow (over days to weeks) during which much of the volume of the lake drained along the valley of the Columbia River to the Pacific Ocean. It is estimated that this type of flooding happened at least 25 times over that period, and in many cases, the rate of outflow was equivalent to the discharge of all of Earth's current rivers combined. The record of these massive floods is preserved in the Channelled Scablands of Idaho, Washington, and Oregon.

Glacial Deposits

Sediments transported and deposited during the Pleistocene glaciations are abundant throughout The United States. They are important sources of construction materials and are valuable as reservoirs for groundwater. Because they are almost all unconsolidated, they have significant implications for mass wasting.

The Bering Glacier is the largest in North America, and although most of it is in Alaska, it flows from an icefield that extends into southwestern Yukon. The surface of the ice is partially, or in some cases completely, covered with rocky debris that has fallen from surrounding steep rock faces. Muddy rivers are distributed from the glacier in several locations, depositing sediment on land, into Vitus Lake, and directly into the ocean. Dirty icebergs are shedding their sediment into the lake. And, not visible in this view, sediments are being moved along beneath the ice.

Figure 19.18 The Bering Glacier in Southeast Alaska. Image by Steven Earle, CC BY 4.0.

The formation and movement of sediments in glacial environments is shown diagrammatically below. There are many types of glacial sediment generally classified by whether they are transported on, within, or beneath the glacial ice. The main types of sediment in a glacial environment are described below.

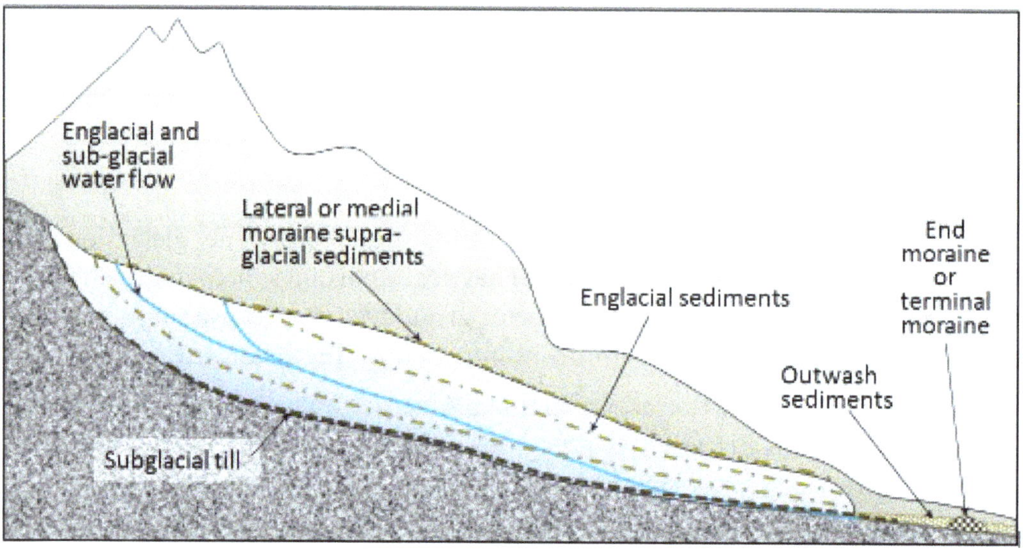

Figure 19.19 A Depiction of the Various Types of Sediments Associated with Glaciation. Image by Steven Earle, CC BY 4.0.

Supraglacial (on top of the ice) and englacial (within the ice) sediments that slide off the melting front of a stationary glacier can form a ridge of unsorted sediments called an end moraine. The end moraine that represents the farthest advance of the glacier is a terminal moraine.Sediments transported and deposited by glacial ice are known as till.

Subglacial sediment, such as lodgement till, is material that has been eroded from the underlying rock by the ice and is moved by the ice. It has a wide range of grain sizes, including a relatively high proportion of silt and clay. The larger clasts (pebbles to boulders in size) tend to become partly rounded by abrasion. When a glacier eventually melts, the lodgement till is exposed as a sheet of well-compacted sediment ranging from several centimeters to many meters in thickness. Lodgement till is normally unbedded. An example is shown in.

Supraglacial sediments are primarily derived from freeze-thaw eroded material that has fallen onto the ice from rocky slopes above. These sediments form lateral moraines and, where two glaciers meet, medial moraines. Most of this material is deposited on the ground when the ice melts and is therefore called ablation till, a mixture of fine and coarse angular rock fragments, with much less sand, silt, and clay than lodgement till. When supraglacial sediments become incorporated into the body of the glacier, they are known as englacial sediments.

Massive amounts of water flow on the surface, within, and at the base of a glacier, even in cold areas and even when the glacier is advancing. Depending on its velocity, this water can move sediments of various sizes and most of that material is washed out of the lower end of the glacier and deposited as outwash sediments. These sediments accumulate in a wide range of environments in the proglacial region (the area in front of a glacier), most in fluvial environments, but some in lakes and the ocean. Glaciofluvial sediments are similar to sediments deposited in normal fluvial environments and are dominated by silt, sand, and gravel. The grains tend to be moderately well rounded, and the sediments have similar sedimentary structures (e.g., bedding, crossbedding, clast imbrication) to those formed by non- glacial streams.

A large proglacial plain of sediment is called a sandur, otherwise known as an outwash plain, and within that area, glaciofluvial deposits can be tens of meters thick. In situations where a glacier is receding, a block of ice might become separated from the main ice sheet and become buried in glaciofluvial sediments. When the ice block eventually melts, a depression forms, known as a kettle, and if this fills with water, it is known as a kettle lake.

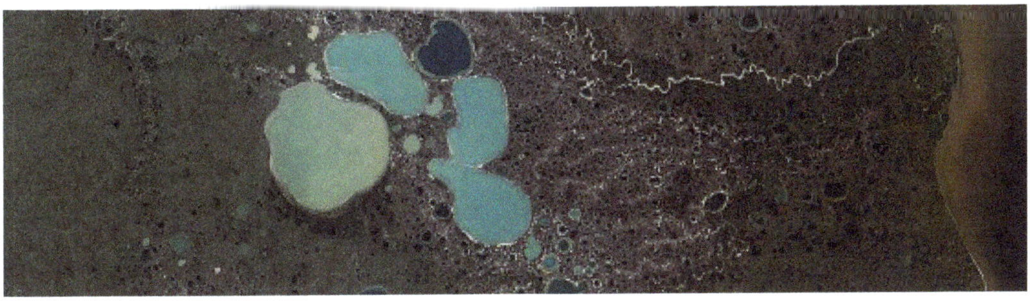

Figure 19.20 Kettle Lakes in The Yamal Peninsula. NASA image by Jesse Allen is in the public domain.

A subglacial stream will create its channel within the ice, and sediments that are being transported and deposited by the stream will build up within that channel. When the ice recedes, the sediment will remain to form a long sinuous ridge known as an esker. Eskers are most common in areas of continental glaciation. They can be several meters high, tens of meters wide, and tens of kilometers long outwash streams commonly flow into proglacial lakes where glaciolacustrine sediments are deposited. These are dominated by silt- and clay-sized particles and are typically laminated on the millimeter scale. In some cases, varves develop; varves are series of beds with distinctive summer and winter layers: relatively coarse in the summer when melt discharge is high, and finer in the winter when discharge is very low. Icebergs are common in proglacial lakes, and most of them contain englacial sediments of various sizes. As the bergs melt, the released clasts sink to the bottom and are incorporated into the glaciolacustrine layers as dropstones.

Figure 19.21 Esker at Fulufjället, Western Sweden. Image by Hanna Lokrantz, CC BY 2.0.

The processes that occur in proglacial lakes can also take place where a glacier terminates in the ocean. The sediments deposited there are called glaciomarine sediments

SUMMARY

There have been many glaciations in Earth's distant past, the oldest known starting around 2,400 Ma. The late Proterozoic "Snowball Earth" glaciations were thought to be sufficiently intense to affect the entire planet. The current glacial period is known as the Pleistocene Glaciation, and while it was much more intense 20,000 years ago than it is now, we are still in the middle of it. The periodicity of the Pleistocene glaciations is related to subtle changes in Earth's orbital characteristics, which are exaggerated by a variety of positive feedback processes.

The two main types of glaciers are continental glaciers, which cover large parts of continents, and alpine glaciers, which occupy mountainous regions. Ice accumulates at higher elevations, above the equilibrium line, where the snow that falls in winter does not all melt in summer. In continental glaciers, ice flows outward from where it is thickest. In alpine glaciers, ice flows downslope. At depth in the glacier ice, flow is by internal deformation, but glaciers that have liquid water at their base can also flow by basal sliding. Crevasses form in the rigid surface ice in places where the lower plastic ice is changing shape. Glaciers are important agents of erosion. Continental glaciers tend to erode the land surface into flat plains, while alpine glaciers create a wide variety of different forms. The key feature of alpine glacial erosion is the U-shaped valley. Arêtes are sharp ridges that form between two valleys, and horns form where a mountain is glacially eroded on at least three sides. Because tributary glaciers do not erode as deeply as main-valley glaciers, hanging valleys exist where the two meet. On a smaller scale, both types of glaciers form drumlins, Roches moutonées, and glacial grooves or striations.

Glacial deposits are quite varied, as materials are transported and deposited in a variety of different ways in a glacial environment. Sediments that are moved and deposited directly by ice are known as till. Glaciofluvial sediments are deposited by glacial streams, either forming eskers or large proglacial plains known as sandurs.

INDEX

M

N

O

P

R

S